T0230856

Principles and Applications of Chemical Defects

Principles and Applications of Chemical Defects

R.J.D. TILLEY
School of Engineering
The University of Cardiff
Cardiff, UK

Routledge
Taylor & Francis Group

LONDON AND NEW YORK

First published 1998 by Stanley Thornes Publishers

Published 2018 by Routledge
2 Park Square, Milton Park, Abingdon, Oxon OX14 4RN
52 Vanderbilt Avenue, New York, NY 10017

First issued in hardback 2018

Routledge is an imprint of the Taylor & Francis Group, an informa business

A catalogue record for this book is available from the British Library.

Typeset and produced by Gray Publishing, Tunbridge Wells, Kent

ISBN 13: 978-1-138-45809-3 (hbk)
ISBN 13: 978-0-7487-3978-3 (pbk)

Contents

Preface

The aim of this book is to provide some insight into chemical defects in crystalline solids. Chemical defects, which are mistakes or changes in the atomic make-up of the crystals, have far-reaching effects on the composition, optical properties and electronic properties of materials and as such is an area of relevance to chemists, physicists, materials scientists and engineers. The book itself has been designed to be read by students with no prior knowledge of the subject, but with a background in basic chemistry and physics. It starts with rather simple ideas but progresses into a discussion of complex materials which are now at the forefront of much intensive research effort.

The book has arisen from courses given to both undergraduate and postgraduate university students in chemistry, physics, materials science and materials engineering. It has built upon the strengths of the previous edition, *Defect Crystal Chemistry*, but has laid additional emphasis on the relationship between basic principles and device applications. This has been accomplished by frequent reference to current research and experimental results and the inclusion of applications throughout the text. The links between principles and applications have been further strengthened by the inclusion of a series of case studies. In addition, the crystal structures that are of most importance have been described throughout the book in a series of boxes, to provide a crystallographic reference within the text.

I have been particularly helped in the revision of this book by E.E.M. Tilley, who first encouraged the idea of revising the text, and by G.J. Tilley and R.D. Tilley, who gave invaluable advice and criticism on the format and contents of the volume. Finally, my indebtedness to my wife Anne continues for her encouragement and tolerance during the rewriting of the text.

R.J.D. Tilley

1 Point defects

1.1 The importance of defects

During the course of this century, and particularly in more recent years, it has been realized that many properties of solids are controlled not so much by the structure of the material itself but by faults or *defects* in this structure. For example, the strength of metals is often governed by the presence of *linear defects* called *dislocations*. Similarly, the various and beautiful colours of many gemstones are due to *impurity* atoms within the otherwise perfect structure of the crystal itself. Impurities are also the key to an understanding of the electronic properties of semiconductors, without which we would have none of the electronic devices that are so important and commonplace today. Defects are also of vital importance in many chemical reactions including *corrosion*, which costs billions of dollars each year, and *catalysis* which generates an equal amount of money by producing essential chemicals for modern industry. Indeed, there is no aspect of the physics and chemistry of solids which is not decisively influenced by the defects that occur in the material.

This book relates to this large subject area and is mainly concerned with the *chemical, optical* and *electronic* consequences of the presence of defects in crystals. The earlier chapters provide an introduction to basic concepts and these are illustrated, where possible, with case studies describing the practical consequences of the ideas that have been presented. In later chapters the concepts are gradually extended, and we will build up the framework of *defect chemistry* and *physics*. This will allow us to be able to meet one of the most important challenges in the modern world, how to manipulate the defect populations in a material so as to endow it with new and desirable properties, that is, *defect engineering*. For example, we will see how, by using our knowledge of defect behaviour, we will be able to accomplish remarkable experimental challenges such as turning a colourless insulator into a black metallic material or a fairly ordinary copper oxide into a high temperature superconductor. Moreover, we will also be able to explain why the defects that we use to achieve these amazing results are able to exert such potent control over the properties of these truly remarkable materials.

At the outset we consider the sort of simple defects that we find in *every* pure crystal of a compound. Even here a surprising number of devices and processes result from the defects present as the two case studies, concerning photography and self-darkening (photochromic) glasses, show.

1.2   Point defects

The simplest *localized defect* that we can imagine in a crystal is a *mistake* at a *single atom site*. These defects are called *point defects*.

1.2.1   Point defects in pure elements

Let us first think of a crystal of a *single element*, diamond, silicon or iron, for example. We can imagine that an atom will sometimes be absent from a normally occupied position. The hole that is left in the structure is called a *vacancy*. Alternatively, during crystal growth an extra atom might be incorporated which is forced to take up a position in the crystal which is not a normally occupied site. These are called *interstitial atoms*. In some cases it is necessary to be a little more explicit in the description and if we want to stress that the interstitial atom is the same as the normal atoms which built the structure, it is called a *self-interstitial atom*. Such vacancies and interstitials, which occur in even the purest of materials, are called *intrinsic* defects. These are shown in Figure 1.1(a) and (b).

Of course, no material is totally pure, and *foreign* atoms will also be present. If these are undesirable or accidental they are known as *impurities*, but if they have been added deliberately, so as to change the properties of the material, they are called *dopant* atoms. The foreign atoms can rest on sites normally occupied by the parent atom type to form *substitutional* defects. Foreign atoms may also occupy positions not normally occupied in the crystal to create *interstitial* impurities or dopants. These defect types are illustrated schematically in Figure 1.2.

At the outset let us say that it is not easy to imagine the three-dimensional consequences of the presence of defects from these two-dimensional diagrams. If it is at all possible try to build crystal models. This will help you to see that in real crystals it will be much easier to create vacancies at some atom sites than others, and that it is easier to introduce interstitials into rather open structures. However, despite any initial difficulties, you will find that this gets easier with practice and you will soon become familiar with the structures involved.

Even though these ideas about defects are not too difficult to follow, they have a great importance. At present almost all of the modern electronics industry, and all microchips that are so widely used in computers and

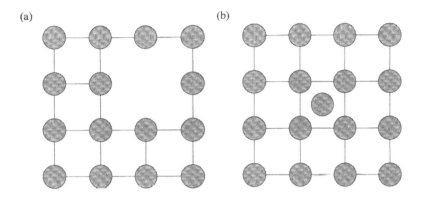

Figure 1.1 A schematic representation of (a) a vacancy and (b) an interstitial, in a monatomic crystal such as silicon or iron.

computer-controlled machinery, are made by the introduction of small amounts of foreign atom dopants into very pure silicon crystals. The dopants are atoms such as phosphorus or aluminium which occupy sites normally occupied by silicon atoms. They are, therefore, substitutional dopants. Later in this book we will find out just how these dopants are able to modify the electronic properties of pure silicon, so that such important devices can be made.

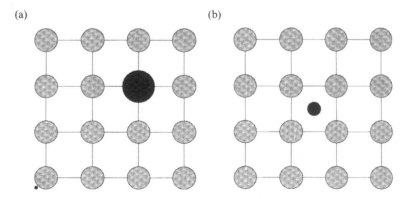

Figure 1.2 A schematic representation of (a) substituted dopant or impurity atom and (b) a dopant or impurity interstitial, in a monatomic crystal such as silicon.

1.2.2   Point defects in compounds

Compounds are made up of more than one atom type. For example, the compound *sodium chloride*, also called *rock salt* or *halite*, shown in Figure 1.3, is composed of equal numbers of sodium (Na) and chlorine (Cl) atoms and has a chemical formula NaCl. In the crystals of any particular compound, the atoms present are always arranged in exactly the same way. This geometrical arrangement is characteristic of the compound and can be used to identify it via X-ray diffraction methods. When we wish to refer to the structure in a generalized way so that we are not really concerned whether the material is, say NaCl itself, but any material with the same arrangement of atoms, then we refer to the *rock salt structure type*.†

It is a good first approximation to regard such materials as being composed of ions, and doing this allows us to extend the ideas outlined about defects. Once again, think of a salt crystal. Reference to Figure 1.3 shows the compound to be built up of an alternation of Na^+ ions and Cl^- ions. We will often want to separate out the effects of the anions from that of the cations and so we will call these two arrays by different names, the *anion sub-lattice* for the Cl^- array and the *cation sub-lattice* for the Na^+ array.

If we now imagine introducing vacancies on the cation sub-lattice the composition and the charge balance will be upset. If x such vacancies occur, the formula of the crystal will now be $Na_{1-x}Cl$ and the overall material will have an excess negative charge of $x-$, because the number of chloride ions is greater than the number of sodium ions by this amount. The compound should be written $[Na_{1-x}Cl]^{x-}$. The same will be true for the anion sub-lattice. If we introduce x vacancies on to the Cl sub-lattice the material will take on an overall positive charge, because the number of sodium ions now outnumbers the chlorine ions, and the formula becomes $[NaCl_{1-x}]^{x+}$. Generally, crystals of salt do not show an overall negative or positive charge or have a formula different to NaCl. So if we imagine vacancy defects in these crystals we need to be sure that the numbers on both the anion and cation sub-lattices are balanced so as to maintain the correct formula and preserve electrical neutrality. This means that we must introduce *equal numbers* of vacancies on to both sub-lattices. Such a situation was envisaged by W. Schottky and C. Wagner, and their ideas were first presented in 1930.

The defects arising from balanced populations of cation and anion vacancies in any crystal, not just NaCl, are now known as *Schottky defects*. For example, if the crystal has a formula MX, then the number of cation

---

†A structure type will be written in italics and a chemical compound in normal type. Hence, *NaCl* refers to the *rock salt* structure type while NaCl means the compound sodium chloride.

Figure 1.3 The NaCl structure. The Na atoms are drawn as small circles and the Cl atoms as large circles.

The rock salt (NaCl) structure

The *rock salt* structure type is adopted by a wide variety of compounds with a formula MX. Some examples are MgO, NiO, CaS and TiN. The unit cell of the *rock salt* structure type is cubic, with a lattice parameter of about 0.5 nm. NaCl itself has a lattice parameter of 0.563 nm. There are four M and four X atoms in a unit cell. Each M is surrounded by an octahedron of X atoms, and each X atom is surrounded by an octahedron of six M atoms.

vacancies will be equal to the number of anion vacancies, in order to maintain electrical neutrality. In such a crystal, one Schottky defect consists of one cation vacancy together with one anion vacancy, although these vacancies are not necessarily imagined to be near to each other in the crystal. It is necessary to remember that the number of Schottky defects in a crystal of formula MX is equal to one half of the total number of vacancies. Schottky defects are frequently represented diagrammatically by a drawing similar to Figure 1.4(a).

In crystals of more complex formula, such as titanium dioxide, TiO_2, there will be twice as many anion vacancies as cation vacancies in a Schottky defect. This is because we need the absence of two O^{2-} ions to electrically counterbalance the loss of one Ti^{4+} ion from the crystal. This ratio of two

(a)

(b)

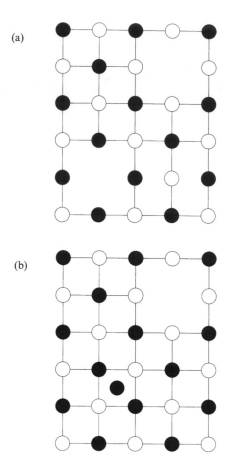

Figure 1.4   (a) Schematic illustration of a Schottky defect on a plane in a crystal of the NaCl type. The defects consist of equal numbers of vacancies on both metal and non-metal sites. (b) Schematic illustration of a Frenkel defect in a crystal of the NaCl type. The defects consist of equal numbers of vacancies on either the metal or non-metal sub-lattice and interstitial ions of the same type. Here the vacancy is on the metal sub-lattice.

anion vacancies per one cation vacancy will hold in all compounds of formula MX_2. In crystals like Al_2O_3, two Al^{3+} vacancies will be balanced by three O^{2-} vacancies. Thus, in crystals with a formula M_2X_3, a Schottky defect will consist of two vacancies on the cation sub-lattice and three vacancies on the anion sub-lattice. As mentioned, these vacancies are not necessarily clustered together and we are only noting the relative numbers needed to keep the crystals electrically neutral.

It is also possible to imagine a defect related to the interstitial defects described above. Such defects are known as *Frenkel defects*, as they were first suggested as being of importance by the Russian physicist Y.I. Frenkel.

In this case an atom from one sub-lattice moves to a normally empty place in the crystal, leaving a vacancy behind. This is shown schematically in Figure 1.4(b). Frenkel defects occur in AgBr. In this compound some of the silver ions move from the normal positions to sit at normally empty places, generating interstitial silver ions, and leave behind vacancies on some of the normally occupied silver sites. The Br^- ions are not involved in the defects at all. We will see later in this chapter that Frenkel defects in AgBr are responsible for the whole of both black and white and colour photography.

In any crystal of formula MX, a Frenkel defect consists of one interstitial atom plus one vacant site in the sub-lattice where that atom would normally be found. Because we are simply moving ions around within the crystal we do not find that we have a problem with charge balance, as we did with Schottky defects. This also means that the relative numbers of vacancies is not connected to the formula of the compound. For example, if we have Frenkel defects on the anion sub-lattice in CaF_2, we can think of just one F^- ion being displaced; it is not necessary to displace two ions to form the Frenkel defect. However, here we can introduce some jargon. Sometimes the term Frenkel defect is reserved for the case when a *cation moves to an interstitial position*, and the term *anti-Frenkel defect* is used for the case where an *anion is displaced.* In this book we will not discriminate in this way and a Frenkel defect may be on either the cation or anion sub-lattice.

Although, Schottky and Frenkel defects do not alter the composition of the host structures, they can have profound effects on the usefulness of materials. It is, therefore, important to try to gain an idea of the numbers of defects present in a normal crystal before we turn to the interesting challenge of how to manipulate these defects to produce new and desirable properties. This forms the basis of the following sections.

1.3 The equilibrium concentration of Schottky defects in crystals

In order to find out if Schottky defects are present in a crystal it is necessary to estimate the energy change that is needed to put the defects into an otherwise perfect solid. This takes us into the realm of thermodynamics, which gives us information about systems when they are at equilibrium. In general, we can say that the first law of thermodynamics tells us that there will be an energy cost in introducing the defects and the second law says that this energy cost may be recouped via the disorder introduced into the crystal. These imprecise concepts can be quantified by using the *Gibbs energy* of a crystal, G, which is written as:

$$G = H - TS$$

where H is the enthalpy, S is the entropy and T the absolute temperature of the crystal. If Schottky defects are going to exist in the crystal, then the

Gibbs energy must be less than it is in a perfect crystal. However, if the introduction of more and more defects causes the Gibbs energy to continually fall, then ultimately there is no crystal left. This means that the form of the Gibbs energy curve must be like Figure 1.5 if defects are to be present at equilibrium. This form of curve turns out to be correct, and we find that Schottky defects exist in *all* crystals at temperatures above 0 K.

In order to determine how many Schottky defects are present in a crystal at equilibrium, we need to estimate the position of the minimum on the Gibbs energy curve shown in Figure 1.5. We proceed in the following way. Introduction of Schottky defects will change the Gibbs energy of the crystal by an amount ΔG, given by

$$\Delta G = \Delta H - T\Delta S$$

where ΔH is the associated change in enthalpy and ΔS the change in the entropy of the crystal. In more simple terms, ΔH is the energy that we must expend in forming the defects and ΔS is the additional randomness in the crystal due to the defects. These two terms correspond to the first law and second law contributions that we mentioned above. To find ΔG, we can try to determine the change in the enthalpy ΔH, and the entropy ΔS, or both. The enthalpy tends to be associated more with nearest neighbours and the bonding energy between them. The change in entropy, ΔS, is also complex, and consists of terms due to the vibration of the atoms around the defects and terms due to the arrangements of the defects in the crystal. Fortunately, this latter quantity, called the *configurational entropy*, is relatively easy to assess using the well-established methods of statistical mechanics, and this term is the one which is estimated. With this information the minimum in

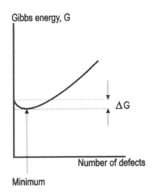

Figure 1.5 A schematic diagram showing the variation of the Gibbs energy needed if a defect population is to be present in a crystal at equilibrium.

the Gibbs energy curve can be obtained by using standard mathematical procedures. The method for a crystal of composition MX is given in Appendix 1.1.

The result is found to be:

$$n_s \approx Ne^{-\Delta H_s/2kT} \tag{1.1}$$

where ΔH_s is the enthalpy to form one defect. The units of ΔH are thus joules per defect and k, Boltzmann's constant, is in JK^{-1}. Sometimes you will find equation (1.1) written in the form:

$$n_s \approx Ne^{-\Delta H_s/2RT} \tag{1.2}$$

In this case ΔH_s is in $J\,mol^{-1}$ and is the energy required to form 1 mole of Schottky defects. This necessitates replacement of Boltzmann's constant by R, the gas constant, which is equal to kN_A where N_A is Avogadro's number, $6.0225 \times 10^{23}\,mol^{-1}$, so that R has units of $J\,K^{-1}\,mol^{-1}$.

It is sometimes useful to make a rough estimate of the *fraction* of sites in a crystal which are vacant due to Schottky disorder rather than compute a more accurate figure from equation (1.2). This figure can be obtained by the following procedure.

Taking logarithms in equation (1.2)

$$\ln n_s = \ln N - \frac{\Delta H_s}{2RT}$$

and as we know that

$$\ln x = 2.3026 \log_{10} x$$

it is possible to write

$$\log_{10} n_s - \log_{10} N = -\frac{\Delta H_s}{(2.3026 \times 2RT)}$$

hence

$$\log_{10}\left(\frac{n_s}{N}\right) = -\frac{\Delta H_s}{4.6052RT}$$

and substituting for R a value of $8.3143\,J\,K^{-1}\,mol^{-1}$ and removing the logarithm term we obtain

$$\frac{n_s}{N} \sim 10^{-\Delta H_s/38T} \tag{1.3}$$

where ΔH_s is measured in $J\,mol^{-1}$.

Remember that this formula only applies to materials with a composition MX, as equation (1.2) was the starting point of the analysis.

Some experimental values for the enthalpy of formation of Schottky defects are given in Table 1.1. We will see in chapters 2 and 3 how these values can be obtained experimentally. Unfortunately the task is not easy and the purity of the crystals studied is of importance. Therefore, there is a large scatter of values in the literature. Those reported in Table 1.1 seem to be among the most reliable available. It is no coincidence that they are for the easily purified alkali halides.

Example 1.1

Obtain a rough estimate of the fraction of Schottky defects present in a crystal at 900 K if the value of ΔH_s is 200 kJ mol^{-1}.

For an approximate answer use equation (1.3)

$$\frac{n_s}{N} \sim 10^{-\Delta H_s/38T}$$

Substituting the values given we find

$$\frac{n_s}{N} \sim 10^{-200\,000/38 \times 900}$$

$$\frac{n_s}{N} \sim 10^{-6}$$

This value shows that the population of point defects is quite low. Even at 900 K we find that only one or two sites in one million are vacant!

Table 1.1   The formation enthalpy of Schottky defects in some alkali halide compounds of formula MX†

Compound	H_s(J) $\times$ 10^{-19}
LiF	3.74
LiCl	3.39
LiBr	2.88
LiI	1.70
NaF	3.87
NaCl	3.75
NaBr	2.75
NaI	2.34
KF	4.35
KCl	4.06
KBr	3.73
KI	2.54

†All compounds have the *rock salt* structure.

Example 1.2

Calculate how the fraction of Schottky defects in a crystal of KCl varies with temperature if the value of ΔH_s is $244 \, kJ \, mol^{-1}$.

To answer this question, we need to compute values of n_s/N from equation (1.2).

$$\frac{n_s}{N} = \exp\left[-\frac{244\,000}{2 \times 8.3143 \times T}\right]$$

substituting suitable values of T (in K!) leads to the values given in Table 1.2.

Example 1.3

Calculate the number of Schottky defects in a crystal of KCl at 800 K.

The fraction of Schottky defects present, n_s/N, can be estimated as above. The problem is to evaluate N. To do this it is necessary to know something of the crystal structure of the material in question. In this case, KCl has the *rock salt* structure. Crystallographic information tells us that the unit cell is a cube of side 0.629 nm and in each unit cell we have four K^+ and four Cl^- ions. The number of sites of each type is then given by

$$N = \left[\frac{4}{(0.629 \times 10^{-9})^3}\right] m^{-3}$$

$$= 1.6 \times 10^{28} \, m^{-3}$$

We can then fill in the details using the data in Table 1.1 to give:

$$n_s = 1.6 \times 10^{28} \exp\left[-\frac{244000}{2 \times 8.3143 \times 800}\right]$$

$$= 1.6 \times 10^{28} \exp[-18.34]$$

Table 1.2　Schottky defect populations in KCl

Temperature (°C)	Temperature (K)	n_s/N
27	300	5.7×10^{-22}
127	500	1.2×10^{-16}
427	700	7.9×10^{-10}
627	900	8.3×10^{-8}

$$= 1.6 \times 10^{28} \times 1.08 \times 10^{-8}$$

$$= 1.7 \times 10^{20} \text{ m}^{-3}$$

1.4   The equilibrium concentration of Frenkel defects in crystals

The calculation of the number of Frenkel defects in a crystal proceeds along lines parallel to those for Schottky defects. We need to find the minimum in the Gibbs free energy curve as a function of defect concentration. To do this we calculate the configurational entropy of the defects in the system as before. However, there is one small difference to take into account. The number of interstitial positions that a displaced ion can move to need not be the same as the number of positions normally occupied and so it is better to make this clear in the calculation. Thus, we suppose that there are N normally occupied lattice sites per m^3 in the array of ions affected by Frenkel defects, and N^* available interstitial sites per m^3 available for the displaced ions to move to. The calculations are set out in detail in Appendix 1.2. It is found that the number of Frenkel defects, n_f, present in a crystal of formula MX at equilibrium is given by:

$$n_f = (NN^*)^{1/2} \, e^{-\Delta H_f/2kT} \qquad (1.4)$$

Values for ΔH_f and Boltzmann's constant k are in $J \, mol^{-1}$ per defect or

$$n_f = (NN^*)^{1/2} \, e^{-\Delta H_f/2RT} \qquad (1.5)$$

when the values and R, the gas constant, are in $J \, mol^{-1}$.

Some experimental values of ΔH_f are given in Table 1.3. As with Schottky defects, it is not easy to determine these values experimentally and there is a large scatter in the values found in the literature. Those given below seem to be among the most reliable.

Table 1.3   The formation enthalpy of Frenkel defects in some compounds of formula MX and MX_2

Compound	H_f (J) $\times 10^{-19}$	Compound	H_f (J) $\times 10^{-19}$
AgCl†	2.32	CaF$_2$‡	4.34
AgBr†	1.81	SrF$_2$‡	2.78
β-AgI†	0.96	BaF$_2$‡	3.06

†Frenkel defects on the cation sub-lattice of a *rock salt* structure compound.
‡Frenkel defects on the anion sub-lattice of a *fluorite* structure compound.

1.5 Schottky and Frenkel defects: trends and further considerations

Perhaps the most important result is that we find that at *all temperatures* above $0\,K$ we should expect to find Schottky or Frenkel point defects present in *all* pure crystals. For this reason such defects are also termed *intrinsic* defects. Moreover, these defects will be in thermodynamic equilibrium, and so it will not be possible to remove them by annealing or other thermal treatments. The actual type of defect found, either Schottky or Frenkel, will mainly depend on the value of ΔH. One would not expect this to be the same for these two alternatives, and would anticipate that only the defect with the lower value of ΔH would be important.

Unfortunately it is not possible to predict, from a knowledge of crystal structure, which defect type will be present in any crystal. Sophisticated calculations, which we discuss in later chapters, show that there is not much difference in the energy required to form either a Schottky or a Frenkel defect in many compounds. However, we can say that often rather close-packed compounds, such as those with a structure like NaCl, tend to contain Schottky defects. The important exceptions are the silver halides, which we will consider in more detail later in this chapter. More open structures, on the other hand, will be more receptive to the presence of Frenkel defects.

Although there is some uncertainty in the experimental values given in Tables 1.1 and 1.3 it is tempting to look for trends in the defect formation enthalpies. However, when doing this it is important to look only among materials which show the same crystal structure type, as a change of crystal structure will usually outweigh the chemical effects. As an example, data for the *rock salt* structure alkali halides is shown in Figure 1.6. This reveals that more energy is required to form defects as we traverse the systems in the direction LiX to KX and that the energy falls as we pass from MF towards MI. However, the differences are not great and it is necessary to be cautious in attributing these trends to any one cause. These factors are best explored via calculations, of which we will say more in chapter 11.

Despite the utility of the formulae given in equations (1.1)–(1.5), they suffer from a number of limitations which it is useful to collect together here.

1. First of all, remember that the formulae derived above apply to materials of formula MX. In order to discuss crystals of different composition, such as M_2X_3, MX_2 and so on, it is important to bear in mind that different, although similar, formulae will result because the configurational entropy term will involve different relative numbers of anion, cation and interstitial sites.
2. Only one sort of defect is supposed to be found in a crystal. This is usually a good assumption to make, but in some systems it has been

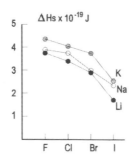

Figure 1.6   The formation enthalpies of Schottky defects in alkali halide crystals.

found that the sort of defect predominating can change, especially at high temperatures. (We will show an example of this in chapter 4.) Close to the transition temperature both defect types will be present.

3. The treatment assumes that the defects do not interact. Now this is not a very good assumption, because if we move ions around in a crystal we would expect electronic interactions to be rather important except when the number of defects present is very small. Defect interactions are important and it is possible to take such interactions into account in more general formulae.

4. The important quantities ΔH and ΔS are assumed to be temperature independent. Once again, this is often quite a good approximation. However, remember that the vibrational component of the entropy, which has been neglected altogether, will become increasingly important at high temperatures.

These three last points show the direction in which the simple theories outlined above can be modified to present a more realistic model of defects in a crystal. The electronic and other interactions between defects can be calculated using a variety of more complex theories, such as the Debye–Hückel treatment used for electrolytes. We will return to this in chapter 11. Different ways of distributing defects over the available lattice positions in a crystal can be envisaged, and ways to estimate the entropy of such distributions can also be sought. This approach can also include more sophisticated *site exclusion rules*, which allow defects to either cluster or keep apart from each other. Nevertheless, our formulae are a very good starting point for our exploration of the role of defects in solids and do apply well when defect concentrations are small and at temperatures which are not too high.

1.6 Case study: the photographic process

1.6.1 Light-sensitive crystals

Photography is one of the most widely used of all information storage methods. Perhaps surprisingly, both black and white and colour photography are possible only because of the presence of point defects in the crystals used. The light-sensitive materials employed in photography are silver halides, notably AgBr, which are dispersed in gelatine to form the photographic emulsion. In order to ensure that the crystals are free of *macroscopic* defects such as dislocations, which degrade the perfection of the photographic images produced, the silver halide crystals are carefully grown within the gelatine matrix itself. The crystals formed are usually thin triangular or hexagonal plates, varying between 0.01 and 10 μm in size, and in photographic parlance are known as *grains*.

When the emulsion is exposed to light a *latent image* is said to form. After illumination each grain will either contain a latent image, that is, it will have interacted with the light photons, or it will have remained unchanged. The film is then put into a developer. Each grain which contains a latent image is totally reduced to metallic silver. Each crystallite with no latent image remains unchanged. The reactions taking place can be written down schematically as:

$$AgBr \rightarrow [AgBr + latent\ image]$$

$$[AgBr + latent\ image] \rightarrow Ag$$

It has been found that only a few photons, maybe as little as six, are needed to form the latent image. As a fully developed crystallite may consist of 10^9 silver atoms, we see that the film is a very sensitive light detector. The final step in the photographic process is called fixing, in which the unreacted silver bromide crystals are removed from the emulsion chemically. These steps are shown schematically in Figure 1.7.

1.6.2 The mechanism of latent image formation

Despite the fact that not all details of the photographic process are completely understood the overall mechanism for the production of the latent image is well known. The halide AgBr, crystallizes in the *NaCl* structure type. Whilst in most crystals with the *NaCl* structure Schottky defects are the major structural point defect type present, it is found that the silver halides, including AgBr, favour Frenkel defects. The formation of latent images is a multi-stage process, involving the Frenkel defect population.

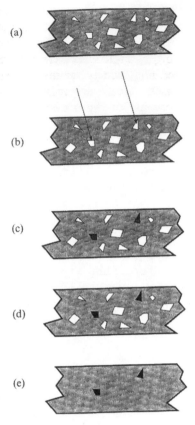

Figure 1.7 The production of an image in a photographic film. (a) The film emulsion contains crystallites of AgBr distributed uniformly in a gelatine layer. (b) Interaction with light introduces latent images into some crystallites, shown grey in (c). (d) Development turns crystallites containing latent images into silver crystals, shown in black. (e) Fixing removes all unreacted AgBr crystals leaving silver crystallites only.

The major steps are believed to be:

1. A light photon interacts with a halogen ion in the AgBr crystal and the energy from the photon liberates an electron from this ion. The liberated electron is free to move in the lattice and migrates to an interstitial silver ion which is part of a Frenkel defect, to form a neutral silver atom.

$$\text{light photon} + \text{Br}^- \longrightarrow \text{e}^- + \text{Br}$$

$$\text{Ag}_i^+ + \text{e}^- \rightarrow \text{Ag}_i$$

where Ag_i^+ represents a silver interstitial ion which is part of a Frenkel defect and Ag_i is a neutral silver interstitial atom.

2. In many instances the above reaction will then take place in the reverse direction, and the silver atom will revert to the normal stable state as a Frenkel defect. However, the metal atom seems to be stabilized if another photon activates a nearby region of the crystal before the decomposition can take place. This stabilization may take place in one of two ways. It is possible that the interstitial silver atom can trap the electron liberated by the second photon to form the unusual Ag_i^- ion, thus:

$$Ag_i + e^- \rightarrow Ag_i^-$$

The silver ion produced in this reaction is then neutralized by association with another interstitial silver atom thus:

$$Ag_i^+ + Ag_i^- \rightarrow 2Ag_i$$

to produce a cluster of two neutral silver atoms.

The second possibility is that the second electron could interact with an interstitial ion to yield a second silver atom which would then diffuse to the first silver atom to form an identical cluster of two:

$$Ag_i^+ + e^- \rightarrow Ag_i$$

$$Ag_i + Ag_i \rightarrow 2Ag_i$$

It is only recently that studies of small metal clusters have shown that a latent image consists of a minimum of four silver atoms. Remarkably, three atoms is not sufficient!

The point which all this leads up to, is that these small clusters of silver atoms completely control the chemistry of photography. To repeat the point stated earlier, when a photographic film is developed, only those crystallites containing a latent image react. The process involves the chemical reduction of the halide to metallic silver. So, the *chemical action* of the developer is *confined to crystallites which contain a latent image*. It becomes apparent that the presence of just one small cluster of four silver atoms determines whether the crystallite can react with developer or not. The point defects that we have recently become acquainted with have quite an importance!

Clearly, successful latent image formation depends on a reasonable concentration of Frenkel defects in the halide crystals. The enthalpy of formation of a Frenkel defect in AgBr is about 2×10^{-19} J and so we can estimate the number of Frenkel defects in a crystal of AgBr at room temperature using the formula given in equation (1.4). The ratio $n_f/(NN^*)^{1/2}$ turns out to be about 10^{-11}. This is not a very high population and it seems reasonable to wonder whether such a concentration is sufficient to allow latent image formation to take place at all. In fact, it is much too small. However, research has shown that the surface of a silver halide grain has a net negative charge which is balanced by an enhanced number of interstitial silver ions located within a few tens of nanometres of the surface. As the halide grains are so small, the ratio of surface to bulk is large, so that the total population of interstitial silver ions in the grains is much higher than the figure given above, which applies to large crystals with relatively small surfaces. This factor is of importance in the overall efficiency of the process and indeed makes practical photography possible.

1.7   Case study: photochromic glasses

Photochromic glass is another material which is sensitive to light. Although many types of photochromic glass have been fabricated, the best known are those which darken on exposure to visible or ultraviolet light and regain their transparency when the light is removed. Such glasses are widely used in sunglasses, sunroofs and for architectural purposes.

The mechanism of the darkening transformation is similar to that involved in the photographic process. Photochromic glasses are complex materials which usually contain silver halides as the light-sensitive medium. The glass for this use would typically be an aluminoborosilicate (Pyrex type) material containing about 0.2 wt% of silver bromide or chloride. In addition, a small amount of a copper chloride is also added. When the glass is first fabricated it is cooled rapidly. Under these conditions the silver and copper halides remain dissolved in the glass matrix and the glass produced is transparent and does not show any photochromic behaviour at all, as shown in Figure 1.8(a) and (b). This glass is transformed into the photochromic state by heating under carefully controlled conditions of temperature and time, which might be, for example, 550 °C for 30 min followed by 650 °C for 30 min. The heat treatment is chosen so that the halides crystallize in the glass matrix, as shown in Figure 1.8(c). Care must be taken to ensure that the crystals do not become too large and that they do not aggregate. A desirable size would be about 10 nm diameter and the individual crystallites should be about 100 nm apart.

It is important that the copper is in the monovalent state and incorporated into the silver halide crystals as an impurity. Because the Cu^+ has the same valence as the Ag^+, some Cu^+ will replace Ag^+ in the AgX crystal, as shown in Figure 1.8(d). Such a crystal is said to be a *solid solution* of CuX in AgX and a new sort of defect has been generated in the crystal, an *impurity defect*. In the present case this consists of a Cu^+ ion occupying an Ag^+ site. These mixed crystallites are precipitated in the complete absence of light. Following this treatment a finished glass blank will look clear because the silver halide grains are so small that they do not scatter light.

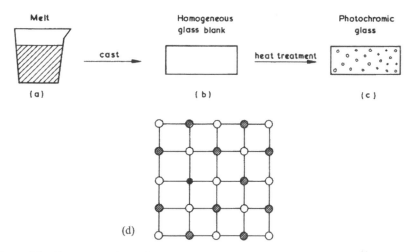

Figure 1.8 The preparation of photochromic glass. (a) A homogeneous melt of glass containing CuCl and AgCl. (b) A non-photochromic glass blank is cast from the melt. (c) Heat treatment transforms the blank into photochromic glass containing crystals of (Ag,Cu)Cl. (d) Schematic illustration of an AgCl crystal containing a Cu^+ impurity ion, shown as a black circle.

The influence of light causes changes similar to those occurring in a photographic emulsion. The photons liberate electrons and these are trapped by interstitial silver ions, which exist as Frenkel defects, to form specks of metallic silver. Unlike the photographic process, the electrons are liberated by the Cu^+ ions which are converted to Cu^{2+} ions in the process.

$$Cu^+ + (light) \rightarrow Cu^{2+} + e^-$$

$$e^- + Ag_i^+ \rightarrow Ag_i$$

$$Ag_i + Ag_i \rightarrow 2Ag_i$$

This process continues until a small speck of silver is created. It is these clusters of silver which absorb the light falling on the glass. The absorption characteristics of the silver specks depend quite critically upon their size and shape. The processes used in producing photochromic glass are manipulated so as to produce a wide variety of shapes and sizes of the silver specks to produce the uniform darkening of the glass.

In a photographic emulsion the halide atoms produced when electrons are released under the influence of light can diffuse away from the silver or react with the emulsion and the process becomes irreversible. In the silver halide crystals, however, the copper ions remain trapped near to the silver particles. This means that the silver particles can release electrons to the Cu^{2+} ions when the light is turned off, making the whole process *reversible*.

This *bleaching process* is the reverse of the darkening process. In fact, the darkening and bleaching reactions are taking place simultaneously under normal circumstances, so that we can speak of *dynamic equilibrium* holding. When the amount of incident light is high we have a large number of silver specks present in the glass and hence a high degree of darkening. When the light intensity falls the number of silver particles decreases and the glass becomes clear again.

As we have seen, photochromic behaviour depends critically upon the interaction of two defects types with light, Frenkel defects in the silver halide together with impurity Cu^{+1} point defects in the silver halide matrix. It is these two defects together which constitute the photochromic phase. For commercially useful materials, the *rate* of the combined reaction is most important. If the darkening takes place too slowly, or if the subsequent fading of the colour is too slow, the materials will not be useful. The presence of the copper halide is essential in ensuring that the *kinetics* of the reaction are appropriate and that the process is reversible.

1.8   Supplementary reading

Frenkel and Schottky defect equilibrium is treated in a number of textbooks, among which are:

W.D. Kingery, H.K. Bowen and D.R. Uhlmann, *Introduction to Ceramics*, 2nd Edition, Wiley–Interscience, New York (1976).
N.N. Greenwood, *Ionic Crystals, Lattice Defects and Non-stoichiometry*, Butterworths, London (1968).
R.A. Swalin, *Thermodynamics of Solids*, 2nd Edition, Wiley–Interscience, New York (1972).

A very clear account, together with self-assessment questions, is given by:

R.F. Davies, *J. Ed. Mod. Mat. Sci. Eng.* **2**, 837 (1980).

Two excellent review articles, which also cover material relevant to chapters 2 and 3 are:

J. Corish and P.W.M. Jacobs, in *Surface and Defect Properties of Solids*, Vol. 2, eds. M.W. Roberts and J.M. Thomas, The Chemical Society, London (1973).
J. Corish, P.W.M. Jacobs and S. Radhakrishna, in *Surface and Defect Properties of Solids*, Vol. 6, eds. M.W. Roberts and J.M. Thomas, The Chemical Society, London (1977).

The photographic process is well documented, and advertising literature often contains much useful information. The classic account of latent image formation is:

R.W. Gurney and N. F. Mott, *Proc. Roy. Soc. Lond., Sect. A* **164**, 485 (1938).

The following are detailed reviews, together with many literature references:

F.C. Brown, in *Treatise on Solid State Chemistry*, Vol. 4, *Reactivity of Solids*, ed. N.B. Hannay, Plenum, New York (1976).
J.F. Hamilton, *Adv. Phys.* **37**, 359 (1988).

Brief but clear accounts of the photographic process and photochromic glasses are given by:

K. Nassau, *The Physics and Chemistry of Colour*, Wiley–Interscience, New York (1983).

The processes occurring in photochromic glasses are described in:

D.M. Trotter, *Sci. Am.* April, 36 (1991).

Detailed information on photochromic and other glass, together with comprehensive bibliographies, are given in:

D.C. Boyd and D.A. Thompson, 'Glass', in *Encyclopaedia of Chemical Technology*, Vol. 11, 3rd Edition, Wiley, New York (1980), 807–880.
R.J. Araujo and N.F. Borrelli, in *Optical Properties of Glass*, eds. D.R. Uhlmann and N.J. Kreidl, Academic Press (1990).

Appendix 1.1
The equilibrium concentration of Schottky defects in crystals

The usual way to tackle this is to start by considering the Gibbs energy of a crystal, G, which is written as

$$G = H - TS$$

where H is the enthalpy, S is the entropy and T is the absolute temperature of the crystal. If we introduce Schottky defects, we introduce a change in the Gibbs energy of the crystal by an amount ΔG, given by

$$\Delta G = \Delta H - T\Delta S$$

where ΔH is the associated change in enthalpy and ΔS the change in the entropy of the crystal. In a crystal of overall composition MX, suppose n_s is the number of Schottky defects per m^3 in the crystal at T K, that is, we have n_s vacant cation sites and n_s vacant anion sites present. In a crystal of this composition there are N possible cation sites and N possible anion sites per m^3. We can determine the entropy change, ΔS, in this system by using the Boltzmann equation:

$$S = k \ln W$$

where S is the entropy of a system in which W is the number of ways of distributing n defects over N sites at random and k is Boltzmann's constant. Probability theory shows that W is given by the formula

$$W = \frac{N!}{(N - n)!n!}$$

where the symbol $N!$, called *factorial N*, is mathematical shorthand for the expression

$$N \times (N-1) \times (N-2)...1$$

Returning to our case, the number of ways that we can distribute the n_s cation and anion vacancies over the available sites in the crystal will be given by the expression

$$w_c = \frac{N!}{(N-n_s)!n_s!}$$

for vacancies on cation sites, and

$$w_a = \frac{N!}{(N-n_s)!n_s!}$$

for vacancies on anion sites. For a crystal of stoichiometry MX

$$w_c = w_a$$

The total number of ways of distributing these defects, W, is given by the product of w_c and w_a, hence

$$W = w_c w_a = w^2$$

Therefore the change in configurational entropy caused by introducing these defects is

$$\Delta S = k \ln(w^2) = 2k \ln w$$

i.e.

$$\Delta S = 2k \ln \left[\frac{N!}{(N-n_s)!n_s!} \right]$$

Now this expression must be simplified somewhat to be of use. We need to eliminate the factorials. This is usually done by employing the approximation

$$\ln N! \approx N \ln N - N$$

which is referred to as Stirling's approximation. In fact, this last approximation is not all that good and is several per cent in error even for values of N as large as 10^{10}. The correct expression for Stirling's approximation is

$$\ln N! \approx N \ln N - N + \frac{1}{2} \ln (2\pi N)$$

which is accurate even for very low values of N. Nevertheless, in order to continue without using excessively cumbersome mathematical expressions we revert to the simpler formula. Substituting, we ultimately obtain

$$\Delta S = 2k\{N\ln N - (N - n_s)\ln(N - n_s) - n_s \ln n_s\}$$

which is (at last!) in a form that we can use.

We make no attempt to calculate the enthalpy change, but merely label the enthalpy needed to form a Schottky defect, ΔH_s. To form n_s pairs we need a total enthalpy input of $n_s \Delta H_s$. The values for ΔS and ΔH_s are substituted into the Gibbs equation to give

$$\Delta G = n_s\Delta H_s - 2kT\{N\ln N - (N - n_s)\ln(N - n_s) - n_s \ln n_s\}$$

In general, the energy increase due to the ΔH_s term will be offset by the energy decrease due to the $-\Delta S$ term. At equilibrium ΔG will be equal to zero and, moreover, the minimum in the ΔG versus n_s curve is given by

$$\left(\frac{d\Delta G}{dn_s}\right)_T = 0$$

i.e.

$$\left(\frac{d\Delta G}{dn_s}\right)_T = \frac{d}{dn_s}\{n_s\Delta H_s - 2kT\left[N\ln N - (N - n_s)\ln(N - n_s) - n_s \ln n_s\right]\} = 0$$

Remembering that $N\ln N$ is constant, so its differential is zero and the differential of $\ln x$ is $1/x$ and of $x\ln x$ is $1 + \ln x$, we find on differentiating

$$\Delta H_s - 2kT\frac{d}{dn_s}[N\ln N - (N - n_s)\ln(N - n_s) - n_s \ln n_s] = 0$$

i.e.

$$\Delta H_s - 2kT\left[\ln(N - n_s) + \frac{(N - n_s)}{(N - n_s)} - \ln n_s - \frac{n_s}{n_s}\right] = 0$$

hence:

$$\Delta H_s = 2kT\ln\left[\frac{(N - n_s)}{n_s}\right]$$

Rearranging:

$$n_s = (N - n_s)e^{-\Delta H_s/2kT}$$

or, if N is considered to be very much greater than n_s,

$$n_s \approx Ne^{-\Delta H_s/2kT}$$

Appendix 1.2
The equilibrium concentration of Frenkel defects in crystals

The calculation of the number of Frenkel defects in a crystal proceeds along lines parallel to those for Schottky defects. Suppose there are N lattice sites per m^3 in the array of atoms affected by Frenkel defects, and N^* available interstitial sites. If n_f ions from the lattice move into interstitial sites, each needing an enthalpy ΔH_f, the total enthalpy change is given by $n_f \Delta H_f$. As before, we turn to an assessment of the configurational entropy of these vacancies and interstitial atoms in order to proceed further.

We can write down the number of ways of distributing the vacancies that have been created over the available positions in the atom array affected by Frenkel defects as

$$w_v = \frac{N!}{(N - n_f)! n_f!}$$

where we have n_f vacancies and a possible total of N positions for the location of the vacancy. Similarly, for the distribution of the interstitial atoms we can write

$$w_i = \frac{N^*!}{(N^* - n_f)! n_f!}$$

where we have n_f interstitials distributed over N^* sites. Proceeding in exactly the same way as for Schottky defects, we can write the total number of ways of arranging the vacancies and interstitials as W, where

$$W = w_v w_i$$

The change in configurational entropy, ΔS, due to this distribution will be given by

$$\Delta S = k \ln W = k \ln w_v w_i$$

Hence

$$\Delta S = k \left\{ \ln \left[\frac{N!}{(N - n_f)! n_f!} \right] + \ln \left[\frac{N^*!}{(N^* - n_f)! n_f!} \right] \right\}$$

Once again we use Stirling's theorem to put this into a more useful format for our needs. This procedure ultimately yields the cumbersome expression

$$\Delta S = k \left[N \ln N + N^* \ln N^* - (N - n_f) \ln (N - n_f) - (N^* - n_f) \ln (N^* - n_f) - 2n_f \ln n_f \right]$$

Note that if we make N^* and N equal to each other we arrive at the expression for Schottky defects. Proceeding as before, the free energy change, ΔG_f, to form the defects, is given by

$$\Delta G_f = n_f \Delta H_f - kT[N \ln N - N^* \ln N^* - (N - n_f)\ln (N - n_f)$$
$$- (N^* - n_f) \ln (N^* - n_f) - 2n_f \ln n_f]$$

Setting

$$\left(\frac{\mathrm{d}\Delta G_f}{\mathrm{d}n_f}\right)_T = 0$$

at equilibrium and differentiating, remembering that $N \ln N$ and $N^* \ln N^*$ are constants and so are, therefore, eliminated, we arrive at an expression similar to the Schottky expression

$$\Delta H_f = kT \ln \left[\frac{(N - n_f)(N^* - n_f)}{n_f^2}\right]$$

i.e.

$$n_f^2 = (N - n_f)(N^* - n_f)e^{-\Delta H_f/kT}$$

This expression can be further simplified if we make yet another approximation, and suppose the number of Frenkel defects, n_f, is a lot less than either the number of normal positions N or the number of interstitial positions available N^*. In this case we can write the approximate expression

$$n_f \approx (NN^*)^{1/2} e^{-\Delta H_f/2kT}$$

2   Atomic mobility: diffusion

2.1   Introduction

The idea of diffusion in a gas or a liquid is well known. If some heavily perfumed flowers are placed in a room the scent is soon noticeable everywhere in the room due to the diffusion of molecules from the flowers through the gas molecules of the air. The diffusion process is tending to make the *concentration* of the 'impurity' perfume equal throughout the volume available. The concentration of the diffusing molecules is greatest at the flowers and least in the furthest corners of the room. This is termed a *concentration gradient* and the diffusion is due to the existence of this concentration gradient. Similarly, if a drop of ink falls into a beaker of water the colour soon spreads out due to the diffusion of the ink particles through the molecules of water. This is also due to a concentration gradient, this time of the ink particles.

Diffusion also takes place in solids, although at a much slower rate than in gases or liquids. However, it remains very important. A piece of clean iron placed outside for several days will soon start to turn 'rusty' and corrode. This is a chemical reaction in which the iron surface reacts with the oxygen in the air to form iron oxide. If we think about this we see that the reaction will stop as soon as a complete film of iron oxide covers the surface unless somehow atoms of iron or atoms of oxygen can traverse the film by *solid-state diffusion* to allow further reactions to occur. Unfortunately, solid-state diffusion is quite easy in iron oxide and so corrosion continues quite rapidly at room temperature. The concentration gradient responsible is set up by the high concentration of oxygen on one side of the oxide film and by the high concentration of iron on the other.

Unlike iron, aluminium metal does not appear to corrode in air. Like iron, initial reaction is the same and a thin film of aluminium oxide forms rapidly on the surface of the clean metal. However, diffusion of aluminium or oxygen across this film is almost impossible and so aluminium appears to be impervious to corrosion under normal conditions.

These differences are extremely important. A great deal of money is spent on trying to prevent iron from rusting, usually with only partial success. It therefore becomes important to ask why these differences should exist. Suppose that you try to think of one atom or ion trying to diffuse through a solid. If all of the atoms in the solid are in their correct lattice sites then the movement can only take place if the diffusing atom uses interstitial positions. That is, the crystals must be able to incorporate some interstitial defects into the structure. It will be easier if there are vacancies present, because the diffusing atoms can then make use of these empty sites to move across the crystal. Thus, we are becoming aware of the fact that solid-state diffusion will be easiest in crystals with large numbers of point defects and most difficult in those with very few point defects.

There are two aspects to solid-state diffusion that need to be described. First, there is the recognition that it is atoms or ions that are moving, that is, we have to think about diffusion at a microscopic or atomic level. However, the effects that can be measured, like the spread of perfume or ink, involve a scale much greater than atoms, a macroscopic level of understanding. The plan adopted in this book has been to discuss the experimentally observable macroscopic effects of diffusion in this chapter. Although we take for granted that atoms move through the solid by one mechanism or another, such mechanistic considerations are put to one side. The following chapter is then reserved exclusively for an interpretation of the experimental results in terms of atom movements. It is here that the importance of point defect populations will be fully appreciated.

Before we can start on this task, it is necessary to define some of the terms common to the subject. Atomic movement through the crystalline lattice is called *volume*, *lattice* or *bulk* diffusion. However, atoms can also diffuse along surfaces or between crystallites. As the regular crystal geometry is disrupted in these regions, atom movement is often much faster than for volume diffusion. Diffusion by way of these pathways is often referred to as *short-circuit* diffusion.

The speed at which atoms or ions move through a solid is usually expressed in terms of a *diffusion coefficient*, which has units of $m^2 s^{-1}$. Not surprisingly, the measured diffusion coefficient of an atom will depend upon a number of factors. The most important of these are: (a) the temperature at which the diffusion occurs; (b) the geometry of the crystal structure through which the atom must move; (c) the number of defects present in the crystal; and (d) whether a chemical reaction takes place as a result of the diffusion.

Because of this it is not correct to talk about *the* diffusion coefficient for an atom but more accurate to define a number of diffusion coefficients, each correct under certain circumstances. The various diffusion coefficients discussed in this chapter are given in Table 2.1. The terms themselves will be explained at appropriate points in the text.

Table 2.1   Symbols and terms for diffusion coefficients

Symbol	Meaning	Applicability
D	Self-diffusion coefficient	Random diffusion processes in the absence of a concentration gradient
D^*	Tracer diffusion coefficient	Diffusion when concentration gradients are small
$\tilde{D}$	Chemical diffusion coefficient	Diffusion in a concentration gradient
$\tilde{D}_{AB}$	Chemical diffusion coefficient	Diffusion coefficient for the reaction between A and B
$\tilde{D}_A$	Interdiffusion coefficient	Diffusion of component A in a chemical diffusion process

2.2   Self-diffusion and tracer diffusion

2.2.1   The determination of tracer diffusion coefficients

When atoms in a pure crystal diffuse under no concentration gradient or other driving force, the process is called *self-diffusion*. In such cases, the atomic movements are random, with motion in one direction just as likely as another and the relevant diffusion coefficient is called the *self-diffusion coefficient* and is given the symbol D.

It is by no means easy, strictly speaking, to measure the self-diffusion coefficient of an atom because it is not possible to keep track of the movements of one atom in a crystal composed of many identical atoms. However, it is possible to measure something which is a very good approximation to the self-diffusion coefficient, if some of the atoms can be uniquely labelled and their movement tracked. In this case the diffusion coefficient that is measured is called the *tracer diffusion coefficient*, written D^*.

There are a number of experimental methods which allow tracer diffusion coefficients to be determined. One common technique uses radioactivity as a means of following the movement of the diffusing atoms. One face of a single crystal is coated with a thin radioactive layer of the same substance. This layer contains the *tracer* atoms. Another slice of crystal is placed on top of the coated crystal, and the sandwich, called a *diffusion couple*, is then heated at a constant temperature for a known period of time. After this treatment, the couple is cut into slices and the radioactivity in each slice measured. This allows the distance moved by the tracer atoms to be determined.

In this experiment there *will* be a concentration gradient, because the concentration of the radioactive isotopes in the coating will be different to the concentration of radioactive isotopes, if any, in the original crystal pieces. This is why the term tracer diffusion coefficient is used. However, if the layer of tracer atoms is very thin, the concentration gradient will be small and will rapidly become smaller as diffusion takes place and in these circumstances D^*, the tracer diffusion coefficient, will be very similar to the self-diffusion coefficient, D.

As an example, to measure the tracer diffusion coefficient of Mg in MgO, which crystallizes with the *rock salt* structure, a thin layer of radioactive Mg is evaporated onto the surface of a carefully polished, single crystal of MgO. This layer is oxidized to MgO by exposing the layers to oxygen gas, after which another carefully polished single crystal slice of MgO is placed on top to form a *diffusion couple*, as shown in Figure 2.1.

The crystal sandwich is heated for a known time at the temperature for which the diffusion coefficient is required. The whole slab is then carefully sliced parallel to the original interface containing the radioactive MgO layer and the radioactivity of each slice, which is a measure of the concentration of radioactive Mg in each section, is determined. A graph of concentration of the radioactive component is then plotted against the distance from the interface to give a *diffusion profile*, or *concentration profile*. The typical form of such profiles is shown in Figure 2.2.

In order to obtain the diffusion coefficient from such profiles, we use a set of equations called *Fick's laws*. These can be applied to a wide range of diffusion problems, including the one to be solved here. For the experiment described, we need to use *Fick's second law* which relates the change in concentration of the diffusing tracer atoms with time to the diffusion coefficient. The experiment described yields information about diffusion in one direction only, perpendicular to the original interface, and so we can use a one-dimensional form of the law. Additionally, if the tracer diffusion coefficient, D^*, is assumed to be independent of the concentration of the radioactive ions, the equation becomes even simpler, and can be written as

$$\frac{\mathrm{d}c}{\mathrm{d}t} = D^* \frac{\mathrm{d}^2 c}{\mathrm{d}x^2} \tag{2.1}$$

where c is the concentration of the diffusing radioactive ions at a distance x from the original interface after time t has elapsed.

(a)   (b)

Figure 2.1 A reaction couple used for diffusion experiments. The slabs are carefully polished and oriented slices of MgO. The central dark strip in (a) represents a very thin layer of MgO containing radioactive Mg atoms. After the diffusion experiment the radioactive Mg has moved away from the original plane, as in (b).

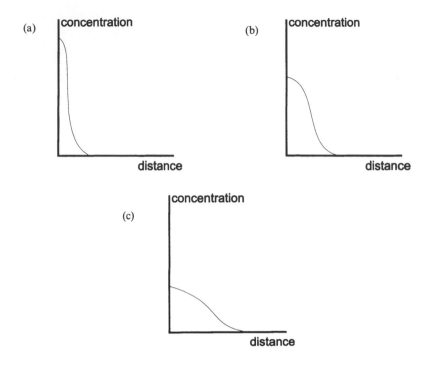

Figure 2.2 Schematic illustration of typical diffusion profiles, which are plots of the concentration of the diffusing species, c, against distance, x, from the original interface. The three curves (a), (b) and (c) refer to three different heating times at the same temperature, (a) being the shortest and (c) the longest.

This equation has to be solved for the experimental arrangement illustrated in Figure 2.1. The answer is found to be†

$$c = \frac{c_0}{2(\pi D^* t)^{\frac{1}{2}}} \exp\left[-\frac{x^2}{4D^* t}\right] \tag{2.2}$$

where c is the concentration of the diffusing species at a distance of x from the original interface after time t has elapsed, D^* is the tracer diffusion coefficient and c_0 is the initial concentration on the surface. A value for the tracer diffusion coefficient is obtained by taking logarithms of both sides of this equation

---

†Different experimental arrangements will result in different solutions to equation (2.1). Fortunately, the appropriate solutions for most geometric and chemical situations of interest have long been derived and are to be found in specialist books on diffusion.

$$\ln c = \ln\left[\frac{c_0}{2(\pi D^* t)^{\frac{1}{2}}}\right] - \frac{x^2}{4D^* t}$$

This has the form

$$\ln c = \text{constant} - \frac{x^2}{4D^* t}$$

and so a plot of ln c versus x^2 will have a gradient of $-1/4D^* t$ as shown in Figure 2.3. A measurement of the gradient gives a value for the tracer diffusion coefficient at the temperature at which the diffusion couple was heated.

It is clear that the experimental procedure must be carried out with care. The crystals must be carefully polished and cleaned before the experiment, and the slices taken after the experiment must be exactly parallel to the surface upon which the radioactive layer was deposited. This allows us to obtain true values of the concentration of the radioactive species as a function of penetration. If we need the diffusion coefficient over a variety of temperatures, as is usually the case, the experiments must be repeated.

2.2.2 Temperature variation of diffusion coefficients

Both tracer and self-diffusion coefficients are usually found to vary considerably with temperature. This variation can often be expressed in terms of the *Arrhenius equation*:

$$D = D_0 \exp\left(\frac{-E}{RT}\right)$$

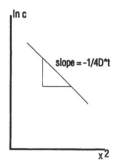

Figure 2.3 A straight line graph of ln c versus x^2, the slope of which can be used to determine the numerical value of the tracer diffusion coefficient, D^*.

named after the Swedish chemist Svante Arrhenius, who formulated this equation to explain kinetic processes in chemical reactions. In this equation R is the gas constant, T is the temperature at which the value of D was measured and D_0 is a constant term referred to as the *pre-exponential factor* or *frequency factor*. The term E is called the *activation energy* of diffusion. It is the energy that is needed to displace a diffusing atom from one stable position in the solid to another.

Taking logarithms of both sides of this equation gives

$$\ln D = \ln D_0 - \frac{E}{RT}$$

If a graph of $\ln D$ versus $1/T$ is drawn, the activation energy can be determined from the gradient of the plot, as shown in Figure 2.4. Such graphs are known as *Arrhenius plots*. Some experimental data, plotted in this way, are presented in Figure 2.5.

Some numerical values of diffusion coefficients will be found in Table 2.2. These data allow trends in the way that diffusion coefficients vary between one compound and another to be picked out. However, because the literature values vary widely this must be done with a certain amount of caution. Despite this qualification we see that structure has a dominant effect. In the metals a change of structure from face-centred cubic (fcc) to body-centred cubic (bcc) increases the pre-exponential factor and decreases the activation energy. This point is well illustrated in the data for iron.

When the *rock salt* structure halides and oxides are considered, the activation energy for diffusion of both cations and anions are rather similar. This is surprising as anions are generally regarded as being much larger than cations. Other factors of importance will become apparent when we consider mechanisms of diffusion in the following chapter.

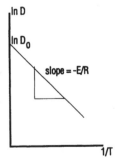

Figure 2.4 An Arrhenius plot of $\ln D$ versus $1/T$, used to determine the activation energy for a diffusion process. The intercept of the line at $1/T = 0$ yields a value for $\ln D_0$ and the gradient yields a value for the activation energy of diffusion.

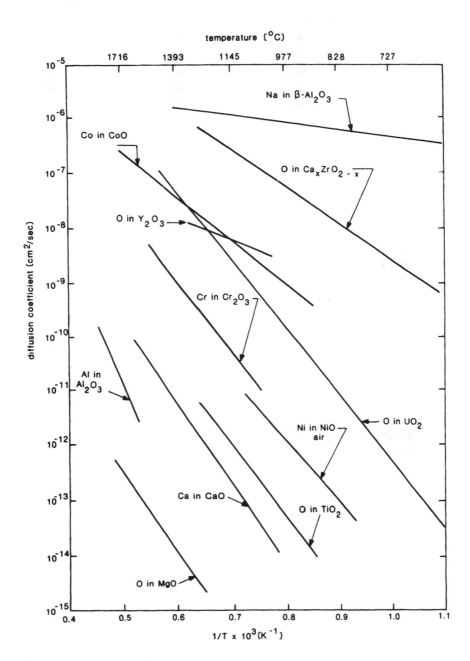

Figure 2.5 Arrhenius plots for diffusion in some common oxides. The y-axis scale is logarithmic, so that values of D are plotted directly against $1/T$. The slope of these graphs yields the activation energy of diffusion.

Table 2.2 Some representative values for self-diffusion coefficients†

Atom	Matrix	D_0 (m²s⁻¹)	E (kJ mol⁻¹)
Metals			
Cu	Cu	2.0×10^{-5}	197
Fe	Fe (fcc)	1.0×10^{-3}	201
Fe	Fe (bcc)	0.5	53
Na	Na	0.16	42
Halides with the *rock salt* structure			
Na⁺	NaCl	8.4×10^{-4}	189
Cl⁻	NaCl	0.167	245
K⁺	KCl	0.5480	256
Cl⁻	KCl	1.3×10^{-2}	231
Oxides with the *rock salt* structure			
Mg²⁺	MgO	2.5×10^{-5}	330
O²⁻	MgO	4.3×10^{-9}	343
Ni²⁺	NiO	4.8×10^{-6}	254
O²⁻	NiO	6.2×10^{-8}	241

†Note: literature values for self-diffusion coefficients vary widely, indicating the difficulty of making reliable measurements. The values here are meant to be representative only.

2.2.3   The effect of impurities

Not all Arrhenius plots are as straightforward as those shown in Figures 2.4 and 2.5. One common form of the graph has two straight line parts, but with differing slopes as shown in Figure 2.6. The point where the two straight lines intersect is called a '*knee*'. If a number of different crystals of the same compound are studied it is often found that the position of the knee varies from one crystal to another. The region corresponding to diffusion at lower temperatures, to the right of the knee, has a smaller activation energy than the region to the left, which normally corresponds to high temperature experiments. The form of the lower temperature region is associated with the impurity content in the crystal whereas the high temperature part reflects the pure material itself. The exact number of impurities in crystals is hard to control and any variation will change the position of the knee. In Figure 2.6, for example, crystal 1 would have a higher impurity concentration than crystal 2. The two parts of the graph are also known as the *intrinsic* region and the *impurity* or *extrinsic* region, respectively. The effect is explained in the following chapter.

2.2.4   The penetration depth

It is quite useful to gain some idea of how far a tracer will diffuse into a solid during a diffusion experiment. This is of considerable practical importance. The electronic properties of integrated circuits are created by the careful

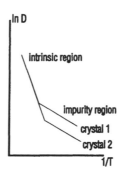

Figure 2.6 A frequently encountered form of an Arrhenius plot. The region at higher temperatures is called the *intrinsic* region, and does not vary greatly from crystal to crystal. The lower temperature curves can occur in a variety of positions, dependent upon the impurity content of the crystals. It is thus referred to as the *impurity* region. In the example shown, crystal 1 would have a higher impurity content than crystal 2.

diffusion of selected dopants into single crystals of very pure silicon. Many metallic machine components are hardened by the diffusion of carbon or nitrogen from the surface into the bulk. In both cases, it is necessary to know the depth to which the diffusing atoms will penetrate.

Now if we think of several single atoms diffusing by a more or less random series of jumps, it is clear that some will penetrate further into the solid than others. Because of this there is no fixed inner boundary to which the tracers diffuse which is why there is no sharp cut-off shown in Figure 2.2. However, it is possible to get an idea of the sort of distances over which the diffusion is appreciable after a certain reaction time in the following way.

In the solution to the diffusion equation (2.2) we see the term $x^2/4D^*t$. This is used to obtain the answer. Generally a quick estimate of the *penetration depth*, x_P, which is the depth where an appreciable change in the concentration of the tracer can be said to have occurred after a diffusion time t, is obtained by equating x_P^2 to D^*t.

$$x_P = (D^*t)^{1/2}$$

In general, this approach is used whenever an estimate is required irrespective of the type of diffusion coefficient available. In chemical diffusion, treated in the following section, we would, therefore, utilize $\tilde{D}$ rather than D^*. In the following chapter we shall gain a more precise idea of the relationship between the penetration depth and the concentration of the diffusing species in the sample.

Example 2.1

The diffusion coefficient of Ni^{2+} tracers in NiO is $10^{-15}\,m^2\,s^{-1}$ at $1100\,°C$. Estimate the penetration depth of the Ni^{2+} ions into a crystal of NiO after heating for $1\,h$ at $1100\,°C$.

The penetration depth is given by

$$x_P = (D^*t)^{1/2}$$

$$= \left(3600 \times 10^{-15}\right)^{1/2}$$

$$= 1.9 \times 10^{-6}\,m$$

2.3   Chemical diffusion

2.3.1   Chemical diffusion coefficients

In the previous section we discussed diffusion in the case where changes in concentration of the diffusing species were *unimportant*. In many systems this is not true, and the measured diffusion coefficient is found to depend upon the concentration of the diffusing atoms. The diffusion coefficient relevant to this situation is called the *chemical diffusion coefficient*, and is written $\tilde{D}$. The chemical diffusion coefficient is related to the concentration and position by a more generalized form of equation (2.1). This will bear upon crystal symmetry and the direction of atom movement, as well as upon the concentration effects that we consider here. For diffusion along only one direction, say x, it would be

$$\frac{dc}{dt} = \frac{d}{dx}\left[\tilde{D}\frac{dc}{dx}\right] \tag{2.3}$$

Unfortunately, it is not possible to solve this equation algebraically because $\tilde{D}$ is not a constant and indirect methods have to be used obtain chemical diffusion coefficients.

In general, reasonably simple experimental conditions hold when one pure metallic element diffuses into another pure metallic element. Because of this, we naturally find that theoretical discussion is often centred around the interdiffusion of two chemically similar metals to form an alloy phase. In order to give a feeling for how diffusion coefficients in chemically reacting systems differ from tracer and self-diffusion coefficients, and how the determination of such diffusion coefficients can be approached in practice, we will stay close to such simple systems in this and the following sections.

Because of the variability of the values obtained experimentally for $\tilde{D}$, it is helpful to try to relate them to self-diffusion coefficients or tracer diffusion coefficients. This objective will form the basis for section 2.4.

2.3.2 The Matano–Boltzmann relationship

One commonly used procedure to obtain numerical values for $\tilde{D}$ is a graphical method described by the *Matano–Boltzmann relationship*. To illustrate this technique, we will consider a case where the diffusion couple consists of two metals, A and B, say copper (Cu) and gold (Au). Slabs of these two metals, ideally single crystal specimens, are placed in contact with one another and heated for an appropriate period of time. During the experiment, some of the metal A will have diffused into metal B and vice versa. After the heat treatment the slabs are carefully sliced up and the variation of the concentrations of A and B are measured across the original interface.

Suppose that the concentration profile after inter diffusion gives the result shown in Figure 2.7(a). To make the discussion as general as possible we will call the concentrations of A which holds at the extreme left of the distribution curve c_A^-, and at the extreme right c_A^+, rather than 0% and 100%. This will prove of use if A and B are present in alloys, say, instead of being present as pure metals. Figure 2.7(b) shows that the curve has been divided by a line, at position x_M, to yield the two shaded areas. In the real diffusion couple, this line will correspond to a plane normal to the x direction. When this plane is chosen so as to make the two shaded areas on Figure 2.7(b) equal to one another, the plane is called the *Matano plane*. If $\tilde{D}$ is the same on both sides of the Matano plane then the curve will be symmetrical with respect to this interface.

Once having drawn the Matano plane we can now determine the chemical diffusion coefficient, $\tilde{D}$, at any value c_A^* that we care to choose, as it is simply given by

$$\tilde{D} = -\frac{\left[\text{area under curve between } c_A^- \text{ and } c_A^*\right]}{\left[\text{slope of curve at } c_A^*\right]2t} \tag{2.4}$$

where t is the time over which the diffusion has occurred and $\tilde{D}$ is valid at the particular value of x^* and the corresponding value of c_A^*. Referring to Figure 2.7(c), we see that the area we need is shaded, and the slope that we need is that of the line labelled S–S in Figure 2.7(d). By repeatedly changing the value of c_A^* and recalculating the new areas and slopes, we can determine how $\tilde{D}$ varies with changing concentration, c_A^*, and with distance from the Matano interface, x^*.

Having given a rather straightforward description of how the chemical discussion coefficient can be evaluated it is useful to write out the *Matano–Boltzmann equation*

$$\tilde{D}_{c_A^*} = -\frac{1}{2t}\left(\frac{dx}{dc}\right)_{c_A^*}\int_{c_A^-}^{c_A^*} x\, dc_A \tag{2.5}$$

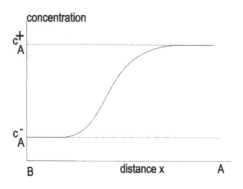

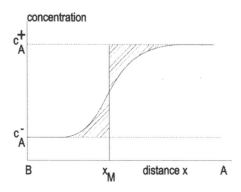

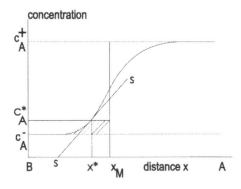

Figure 2.7 (a) A measured penetration curve for component A after inter-diffusion of A and B. (b) The Matano plane, at x_M, is placed so that the shaded areas are equal. (c) The area to be measured, shaded, and the slope of the curve needed, the line S–S, in the Matano procedure is for the evaluation of the chemical diffusion coefficient at concentration c_A^* and position x^*.

This expresses in mathematical terms the verbal equation (2.4). The integral in equation (2.5) represents the shaded area in Figure 2.7(c) and the denominator in equation (2.5) represents the slope of the curve at c_A^* in Figure 2.7(d).

The diffusion coefficient that is found from the Matano–Boltzmann analysis is the single diffusion coefficient that describes the reaction between A and B and not the diffusion coefficients for either of the components separately. It is analogous to the voltage given by a battery, which is the sum of the voltages due to the cathode and the anode and not just the voltage from one or the other electrode alone. For this reason it is best to write it as $\tilde{D}_{AB}$.

2.4 Chemical diffusion, intrinsic diffusion and self-diffusion

2.4.1 The Kirkendall effect

In the preceding section we used the Matano interface as a reference plane when it was necessary to measure distances in a diffusion couple. Intuitively it seems far simpler to use the initial interface between the two reactants, A and B, for this purpose. This is a nice idea, but unfortunately it is not always easy to locate this plane after diffusion has occurred.

Let us think about this a little. After heat treatment, the central region of the diffusion couple will consist of a new material, called the *inter-diffusion phase*. If we suppose that A diffuses twice as quickly across the interface into B, as B does in the opposite direction, then clearly the amount of inter-diffusion phase on one side of the interface will be twice as much as on the other side, and the diffusion coefficient, or at least the relative diffusion coefficients, can easily be determined by simple measurement. So, why not mark the initial interface? We could do this by placing a few inert markers, platinum wires, for example, at the interface before the heating cycle is started. Unfortunately, the inert markers will not always remain in place at the interface.

This shift of markers is known as the *Kirkendall effect*. It was first observed in an experiment in which a block of α-brass (70% Cu: 30% Zn) was embedded in a block of copper. The brass was wrapped around with fine molybdenum wires which were to act as the inert marker, as shown in Figure 2.8(a). After heating, it was found that the separation of the wires had decreased.

The reason for movement of the marker is not too difficult to understand. Assume that markers are placed at the interface between the components, as shown in Figure 2.8(b), that the diffusion of A is faster than the diffusion of B and that the volume of the system after the diffusion experiment is the same. That is, alloy formation does not alter the total volume of the couple.

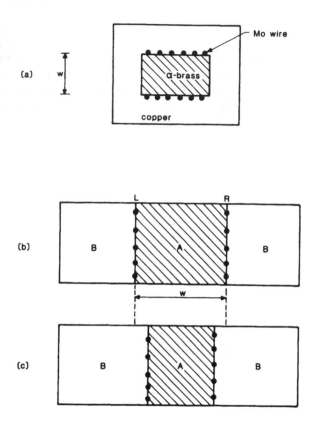

Figure 2.8 (a) Schematic illustration of the Kirkendall effect. (b) The situation in (a) before the reaction. (c) The situation in (a) after the reaction, if component *A* diffuses faster than component *B*. The separation of the markers, *w*, appears to decrease.

After some time, the number of atoms of *A* which has passed to the right of the marker R is greater than the number of atoms of *B* which has passed to the left of R. This is shown in Figure 2.8(c), where we now see that the amount of material to the left of the marker is smaller than the amount of material to the right. The opposite happens at the marker L. Clearly, we started out with equal volumes on both sides of the marker, and so it appears, in an experimental observation, that the marker R has *moved to the left* and marker L has *moved to the right*. The separation between the markers, *w*, thus appears to decrease.

In reality, the extent of the Kirkendall effect is difficult to estimate. There will invariably be some volume change because the molar volume of the alloy formed will be different to that of either of the initial components, although this may be small. In addition, if one component leaves one area of sample faster than the other component moves in, we are likely to get voids formed. When this happens it also effects the marker displacement. It is, therefore, difficult to treat the Kirkendall effect theoretically in a general way. One situation has been analysed in detail, and we shall briefly discuss this now, as it will allow us to separate the contribution of the diffusion coefficient of component A from that of component B in the reaction.

2.4.2 Intrinsic diffusion coefficients

The chemical diffusion coefficient, $\tilde{D}$, for our A–B system can be thought of as made up of two *intrinsic diffusion coefficients*, $\tilde{D}_A$ and $\tilde{D}_B$. These are equivalent to the chemical diffusion coefficients of each of the separate components A and B in our reaction. In general, $\tilde{D}_A$ and $\tilde{D}_B$ as well as $\tilde{D}$, are concentration dependent. It is, therefore, not possible to relate these quantities to each other over all the concentration ranges that apply in the diffusion couple, but in the moving marker plane, also called the *Kirkendall plane*, some relationships between them can be found.

This is not simply an arbitrary choice of origin, of course. Suppose, for example, we wish to measure the diffusion coefficient of a dye molecule in water. The dye could simply be injected into a tank of water at a certain marked spot, and its spread observed visually. Unfortunately, in the case of chemical diffusion, it is rather like trying to do the experiment by dropping the dye into a river rather than a tank of still water. In this case two factors are operating. Firstly, the dye is diffusing outwards, as before, but also the whole body of water in which the dye diffuses is being swept along downstream. To study the diffusion process alone it is necessary to walk downstream at the same speed as the flow and then the spread of the dye can be measured without introducing an error due to the water flow. To put this into more formal language, the point in the river where the dye is injected is taken as a *moving* reference point. Using the moving Kirkendall marker plane is the analogous situation in our diffusion couple.

Bearing this in mind, we find, not surprisingly, that the movement of the marker plane with respect to the Matano plane is quite simply related to the *difference* in the separate diffusion coefficients, $(\tilde{D}_A - \tilde{D}_B)$. The key equation is

$$vel_K = \left(\tilde{D}_A - \tilde{D}_B\right)\left(\frac{\mathrm{d}N_A}{\mathrm{d}x}\right) \tag{2.6}$$

where vel_K is the velocity of the marker plane, N_A is the mole fraction† of component A. The other key equation is

$$\tilde{D} = N_A\tilde{D}_B + N_B\tilde{D}_A \tag{2.7}$$

where N_B the mole fraction of component B. Equation (2.7) is known as the *Darken equation*. In both equation (2.6) and (2.7) measurements are made at the marker plane.

These equations allow values for $\tilde{D}_A$ and $\tilde{D}_B$ appropriate to the Kirkendall plane to be determined. The following procedure is employed. First, $\tilde{D}$ and the mole fractions of A and B are measured in the Kirkendall plane using the Matano–Boltzmann method. This gives information on $(\tilde{D}_A + \tilde{D}_B)$ from equation (2.7). Then the slope of the curve is measured at the Kirkendall plane, which gives us dN_A/dx. Finally, the displacement of the Kirkendall plane from the Matano plane, x_K, is measured and as the time of the diffusion experiment t, is known, we can calculate the velocity of the marker plane, vel_K, using the equation:

$$vel_K = \frac{x_K}{2t}$$

We now have sufficient information to calculate separate values for both $\tilde{D}_A$ and $\tilde{D}_B$ in the Kirkendall plane. If the marker plane coincides with the Matano plane then vel_K will be zero and so $\tilde{D}_A$ will be equal to $\tilde{D}_B$.

2.4.3 The relationship between chemical diffusion and self-diffusion coefficients

It is difficult to obtain a general relationship between chemical diffusion and self-diffusion coefficients, but, as above, it is possible to obtain a relationship which is valid in the in the Kirkendall plane. For component A

$$\tilde{D}_A = D_A^* F$$

where $\tilde{D}_A$ is the intrinsic diffusion coefficient and D_A^* the tracer diffusion coefficient of component A, and F is called the *thermodynamic coefficient*. A similar equation can be written for component B. Therefore, the chemical diffusion coefficient can be regarded as the tracer diffusion coefficient multiplied by a factor, the thermodynamic coefficient, which accounts for concentration changes. At low concentrations, or when concentration

---

†The mole fraction of A is the number of moles of A divided by the total number of moles of A and B present,

$$N_A = \frac{\text{moles of } A}{\text{moles of } A + \text{moles of } B}$$

The number of moles of A is equal to the mass of A present, in grammes, divided by the molar mass of A, in grammes.

$$\text{moles of } A = \frac{\text{mass of } A \text{ (g)}}{\text{molar mass of } A \text{ (g)}}$$

effects are not relevant, the value of F is unity and $\tilde{D}_A = D_A^*$. For chemically similar metals or compounds it is often possible to take the thermodynamic coefficient for component B to be the same as that for component A, which allows us to write

$$\frac{\tilde{D}_A}{\tilde{D}_B} = \frac{D_A^* F_A}{D_B^* F_B} = \frac{D_A^*}{D_B^*}$$

These relationships between the chemical diffusion coefficient and the tracer diffusion coefficient are called the *Darken relations*. In general, they are found to hold well.

2.5   Diffusion in ionic crystals

2.5.1   Ambipolar diffusion

In the previous sections of this chapter diffusion was restricted to neutral atoms and was considered to be due to changes in concentration. When the movement of charged particles, ions or electrons, is considered, concentration is not the only factor of importance. In considering the diffusion of charged particles, which is referred to as *ambipolar diffusion*, overall electric charge neutrality *must be maintained* during diffusion. In order to examine this concept in a little more detail it is instructive to consider some typical experimental situations.

2.5.2   Solid solution formation

If two compounds with the same formula and crystal structure are placed in contact, the atoms in each can inter-diffuse to form a *mixed crystal* or *solid solution*. In the case of ionic compounds such as oxides, this is achieved by the diffusion of ions. A concrete example is provided by the reaction between NiO and MgO to form a mixed crystal of composition $Ni_xMg_{1-x}O$. Both of these oxides crystallize with the *rock salt* crystal structure. This means that the oxygen anion sub-lattice is the same in both crystals. Thus, if we ignore any slight rearrangements of the anions at the interface between the two crystals, the solid solution formation will involve only the diffusion of Mg^{2+} and Ni^{2+} cations in opposite directions.†

The situation is shown in Figure 2.9. An experimentally determined diffusion profile, obtained by heating a single crystal of NiO placed in contact with a single crystal of MgO, is shown in Figure 2.10. Note that the

---

†Note that self-diffusion of the oxygen atoms will occur but this will not contribute to the formation of the solid solution.

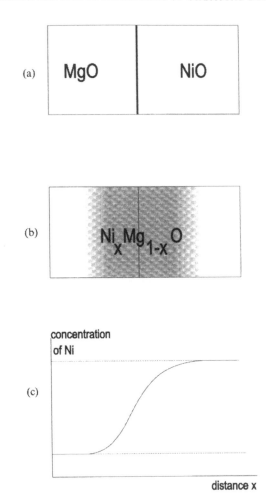

Figure 2.9   A diffusion couple in which a solid solution forms between the starting phases MgO and NiO. (a) The situation before reaction. (b) The situation after the reaction. Note that the boundaries between both MgO and NiO and the solid solution will be diffuse. (c) Plot of the concentration of Ni versus distance across the couple.

interface is not symmetrical and so the diffusion coefficients of the two cations must differ.

Now this observation raises a problem. If one of these ions diffuses faster than the other then, after some time, on one side of the boundary there will be fewer cations present and on the other side more cations present, than at the outset. This will create charge balance difficulties. In the NiO–MgO system a small concentration of mobile electrons is present because of the slightly non-stoichiometric behaviour of NiO, as explained in chapters 6 and

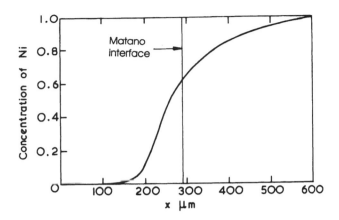

Figure 2.10 An experimentally determined concentration profile for the system NiO–MgO heated at 1370 °C in air. Note that the curve is unsymmetrical. [Redrawn from S.L. Blank and J.A. Pask, *J. Am. Ceram. Soc.* **52**, 669 (1969).]

8. These electrons always allow local charge balance to be maintained, as they migrate very quickly compared to the ions themselves, so that electric field effects do not present problems.

Example 2.2 Estimate the chemical diffusion coefficient applicable to the reaction shown in Fig. 2.10

The experimental data can be analysed by the Matano procedure to obtain the diffusion coefficient as a function of the distance from the Matano interface and the concentration of Ni^{2+}. At the temperature of the experiment shown in Figure 2.10, 1370 °C, it was found that $\tilde{D} = 1.25 \times 10^{-15}\,m^2\,s^{-1}$ at a nickel concentration of 10 at% and $6.5 \times 10^{-15}\,m^2\,s^{-1}$ at a nickel concentration of 40 at%.

2.5.3 Spinel formation

A slightly different diffusion problem occurs if an intermediate phase is formed instead of a solid solution when two compounds are put into contact. The *mechanism of the reaction* may depend upon whether electron transport is possible in the intermediate phase and the *rate of reaction* will be controlled by the rate of diffusion of the slowest ion.

A good example of this situation is provided by spinel formation. Spinel is a mineral with a composition $MgAl_2O_4$. A large number of other oxides crystallize with the same structure type and formula AB_2O_4. These are collectively referred to as *spinels*. Most often A represents a divalent cation and B a trivalent cation, as in $MgAl_2O_4$ itself, but other combinations of ions are also possible. The spinel structure is discussed in detail later in this book, but for the present this information is not needed.

The spinel formation reaction can be represented by the chemical equation

$$MgO + Al_2O_3 \longrightarrow MgAl_2O_4$$

Suppose that an experiment similar to solid solution formation is set up, but this time using a crystal of Al_2O_3 in contact with a crystal of MgO, as shown in Figure 2.11(a). The initial reaction will result in the separation of the two reacting oxides MgO and Al_2O_3 by a layer of spinel, $MgAl_2O_4$. Continued reaction will depend upon transport of reactants across the spinel layer. A number of mechanisms can be suggested but, because $MgAl_2O_4$ is an insulator, electron transport is not possible and so only mechanisms involving ions are permitted. These are shown in Figure 2.11(b)–(d).

In Figure 2.11(b) the reaction is sustained by diffusion of equal numbers of O^{2-} anions and Mg^{2+} cations. The electrical charges on the ions are equal and opposite so no charge balance problems arise. Because the ionic movement is towards the right, new spinel growth will take place at the right-hand side of the spinel layer. In Figure 2.11(c) the reverse situation is shown where diffusion of O^{2-} anions is accompanied by a parallel diffusion of Al^{3+} cations. Because of the difference in the ionic charges, two Al^{3+} cations need to be accompanied by three O^{2-} anions to maintain charge neutrality, as shown. Spinel growth will now take place at the left-hand boundary. Finally, Figure 2.11(d) shows a scheme in which only cations diffuse, Mg^{2+} from left to right and Al^{3+} from right to left. In order to maintain the charge balance, for every three Mg^{2+} cations which diffuse in one direction two Al^{3+} must move in the other. In this case, the spinel layer forms on either side of the initial boundary.

It has been found that the reaction between MgO and Al_2O_3 follows the mechanism shown in Figure 2.11(d). At the boundary between Al_2O_3 and spinel we have the reactions

$$Al_2O_3 \longrightarrow 2Al^{3+} + 3O^{2-}$$

$$3Mg^{2+} + 3O^{2-} + 3Al_2O_3 \longrightarrow 3MgAl_2O_4$$

While at the boundary between MgO and spinel we have the reactions

$$3MgO \longrightarrow 3Mg^{2+} + 3O^{2-}$$

$$3O^{2-} + 2Al^{3+} + MgO \longrightarrow MgAl_2O_4$$

These equations indicate that the spinel layer grows in an asymmetrical fashion. For every three Mg^{2+} ions which arrive at the Al_2O_3 boundary

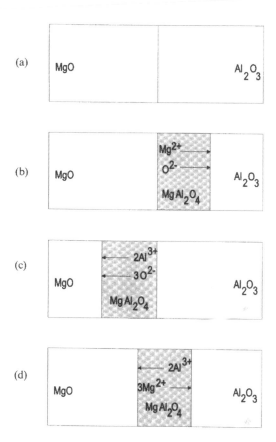

Figure 2.11 Schematic illustration of some diffusion processes which can occur in spinel formation between two oxides MgO and Al_2O_3. (a) Diffusion of Mg^{2+} cations and O^{2-} anions allows reaction to take place at the right-hand surface of the spinel layer. (b) Diffusion of Al^{3+} cations and O^{2-} anions allows the reaction to take place at the left-hand surface of the spinel layer. (c) Counter diffusion of both cations allows the reaction to take place at both surfaces.

three $MgAl_2O_4$ molecules form, while for every two Al^{3+} ions which arrive at the MgO boundary only one $MgAl_2O_4$ molecule forms. Therefore, the spinel layer will form in a ratio of 1:3 on either side of the initial boundary, with the thicker part on the Al_2O_3 side.

The rate at which the total thickness of the spinel layer grows is controlled by the speed of diffusion of the slowest cation. In reactions of this sort, if the spinel layer increases by an amount Δx in a period of time Δt, the rate of film growth, $d(\Delta x)/d(\Delta t)$, is given by

$$\frac{d(\Delta x)}{d(\Delta t)} = \frac{k}{x}$$

where k is a constant and x is the film thickness at time t. Integration and rearrangement of this equation leads to

$$x^2 = 2kt$$

where x is the thickness of the *spinel* layer, k is called the *practical reaction rate constant* and t is the reaction time. Because a graph of x versus t is parabolic in shape, k is also sometimes called the *parabolic rate constant*.

2.6 Case study: corrosion and oxidation reactions

Corrosion occurs when a metal is attacked by a gaseous atmosphere, which is damp oxygen in everyday life, to produce a thin layer of product phase. It has been observed since antiquity that the 'Noble' metals, especially gold, were those which did not appear to corrode in air. Owing to the industrial and economic importance of such reactions, they have been very extensively studied. It was soon appreciated that in order for the corrosion to continue some of the reactants must diffuse across the product layer. For many metals, such as transition metals, this is not too difficult, even at room temperature. For others, like aluminium, diffusion is not possible. This means that in practice aluminium appears to be resistant to corrosion, although in effect it is the thin aluminium oxide film which quickly forms on the metal which is the protective barrier. Such resistance to diffusion is also found in the lanthanide metals. Lighter 'flints' are made from a mixture of lanthanide metals, known as *mischmetall*. The metal of the flint is protected by a thin oxide film which prevents further oxidation. When the flint is abraded against a steel surface, as when using the lighter, the oxide film is removed and the subsequent reoxidation of the newly exposed metal generates enough heat to cause sparks.

The reasons why some metal oxides can support diffusion, while others do not, is explained later in this book. In this section the role that diffusion plays in corrosion reactions will be explored.

The classical studies on metal oxidation involved the formation of Cu_2O. When copper metal is oxidized in conditions of low partial pressure of oxygen a thin film of Cu_2O forms. The reaction is described by the chemical equation

$$2Cu + \frac{1}{2}O_2 \rightarrow Cu_2O$$

Because of this we will focus on the situation that arises when the initial reaction results in the formation of a continuous film of product which is firmly attached to the metal surface. In such a case, further reaction can only proceed if ions or atoms can diffuse across the product, either from the outside gaseous phase into the inner metal layer, or else from the metal out to meet the gas. If this cannot happen then no reaction takes place.

There are several possible mechanisms for the transport of material across the initial oxide coating. These are shown in Figure 2.12. Because oxides are rather ionic compounds it is usual to think of ions as the moving species and not copper atoms or oxygen molecules. Perhaps the most obvious possibility, illustrated in Figure 2.12(a), employs the diffusion of Cu^+ cations outward from the metal towards the gas atmosphere. If this took place, a large negative charge would be left behind at the copper–copper oxide interface which would soon slow down the moving cations and bring the reaction to a halt. To maintain electrical neutrality in the system and to allow the reaction to continue, this diffusion must be accompanied by a parallel diffusion of an equal number of electrons. When the electrons arrive at the surface they react with oxygen molecules on the oxide surface to form two O^{2-} ions. These are incorporated into the oxide film and together with the arriving Cu^+ cations allows the film to grow from the outer surface of the film.

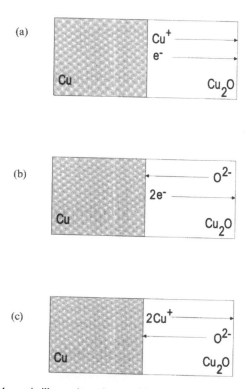

Figure 2.12 Schematic illustration of some diffusion processes which can allow the continued oxidation of copper (Cu) to Cu_2O. (a) Diffusion of Cu^+ cations and electrons allows further reaction to take place at the outer surface of the oxide. (b) Counter diffusion of O^{2-} anions and electrons allows the reaction to take place at the inner surface of the oxide. (c) Counter diffusion of both anions and cations allows the reaction to take place at both surfaces.

The scheme illustrated in Figure 2.12(b) shows the diffusion of oxygen ions into the film. These ions cannot be generated spontaneously, and it is necessary for electrons to move through the oxide layer from the copper to make the ionization possible. This leaves Cu^+ cations behind at the copper–copper oxide boundary. These cations are able to combine with the arriving O^{2-} anions to extend the oxide film at the copper–copper oxide boundary.

In the third scheme, shown in Figure 2.12(c), the mechanism involves the counter diffusion of Cu^+ cations and O^{2-} anions. On the face of it, this mechanism does not need electron movement to proceed. However, anions cannot be continually generated at the surface nor cations at the metal–metal oxide interface without some electron movement. Fortunately, it is not necessary to worry about these electrons, or the electrons in the other reaction schemes, at this stage. This is because the electrons move much faster than the ions and so the rate of corrosion is controlled by ionic diffusion alone.

Experiments have shown that the mechanism shown in Figure 2.12(a), involving diffusion of Cu^+ cations, is followed. This is largely because the O^{2-} ions are too large to diffuse readily through the structure, which eliminates both of the other reactions shown.

The rate of growth of the film is controlled by the slow diffusion of the Cu^+ ions. It is quite easy to determine this rate by measuring the gain in weight of a copper strip over time. To do this, a strip of copper is suspended from a sensitive balance and enclosed so that the oxygen pressure surrounding the strip can be varied. The copper strip is then heated at a suitable temperature and the weight gain recorded. In reactions controlled by diffusion, the rate of thickening of the film obeys the parabolic rate law described for *spinel* formation, above. The increase in the thickness of the film, x, varies with time in the following way

$$x^2 = 2k_p t$$

where k_p is called the *parabolic rate constant* for the formation of a layer of Cu_2O of thickness x after reaction time t.

The mobile electrons play an important role in this reaction. Although the rate of the reaction is controlled by the slow cation diffusion, in the absence of any mobile electrons the reaction would stop. Indeed, this is the reason why both aluminium and lanthanide metals do not obviously corrode. As we will see in later chapters, the ability of these oxide films to support electronic conduction is often a reflection of the ability of the cation to adopt more than one valence state. This is a characteristic of transition metal compounds and so these are particularly susceptible to corrosion.

2.7 Short-circuit diffusion

In the preceding sections we have focused attention upon diffusion through a crystal lattice which contained only point defects. However, diffusion along other imperfections in the structure is often much faster than bulk diffusion, especially at lower temperatures. This process is referred to as *short-circuit diffusion*. The main imperfections that are of importance for short-circuit diffusion are *dislocations* and *grain boundaries*. Dislocations can be thought of as *tubes* of disordered structure which thread through the crystal, and grain boundaries are the *surfaces* between individual crystallites in a polycrystalline solid.

The situation which has been studied in most detail is that of diffusion along a grain boundary. One way of experimentally determining the effect of grain boundary diffusion is to follow a route very similar to that described in section 2.2 for the determination of tracer diffusion coefficients. Initially a polycrystalline material is carefully polished and a thin layer of a radioactive tracer is then coated onto the polished surface, as in Figure 2.13. The sample is now heated at an appropriate temperature for some hours during which time the tracer will diffuse into the material.

Several processes are now occurring simultaneously. The tracer will diffuse into the crystal lattice from the surface at a speed related to the tracer diffusion coefficient. It will also diffuse down the grain boundary at a rate characteristic of the grain boundary diffusion coefficient. Finally, the tracer will also diffuse sideways into the bulk from the grain boundary, again at a rate characteristic of the tracer diffusion coefficient. The contours of equal concentration of the diffusing tracer in the polycrystalline material will depend upon the relative values of the diffusion coefficient in the bulk compared to that down the grain boundary. If the grain boundary penetration is much greater than the penetration into the crystal lattice, a profile as shown in Figure 2.13(b) will result. If the rate of movement along the grain boundary is only a little faster than that through the lattice not much effect will be noticed. This is shown schematically in Figure 2.13(c).

After the experiment, the sample is carefully sliced parallel to the top surface and the radioactivity measured, which allows a penetration profile to be drawn which is analogous to that shown in Figure 2.2. However, this only reveals part of the information about the process that could be obtained. In order to obtain some details about the distribution of the radioactivity a piece of photographic film can be placed in contact with the freshly cut surface. The radiation from the tracer effects the photographic film in the same way as light photons. Thus, the darkening of the film is proportional to the degree of radioactivity in the underlying sample and gives a nice picture of the distribution of the tracer across the exposed surface.

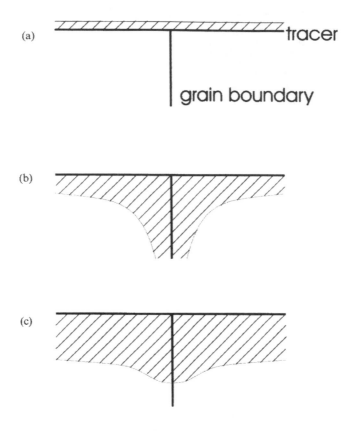

Figure 2.13 Short-circuit diffusion down a grain boundary. (a) The situation before reaction showing a surface layer of tracer atoms. (b) Diffusion of the tracer is much faster down the boundary than in the bulk. (c) Diffusion of the tracer is hardly faster down the boundary than in the bulk. The shaded areas in (b) and (c) indicate the penetration depth of the tracer.

The sort of results obtained by the combination of techniques illustrated in Figure 2.14. The penetration curve is shown as a line, clearly divided into two segments. The photographic 'snapshots' are shown as insets labelled from (a) to (f). In the surface layers, the radioactive tracer will have saturated the lattice and the grain boundaries, the average radioactivity will be high and it will not be possible to distinguish the grain boundaries from the surrounding crystal, as shown schematically in Figure 2.14(a) and (b). However, as the crystal is sliced the average radioactivity will fall and the grain boundaries start to show higher levels of radioactivity than the surrounding lattice, as shown in Figure 2.14(c) and (d). Ultimately the radioactivity is only associated with the grain boundaries, as seen in Figure 2.14 (e) and (f).

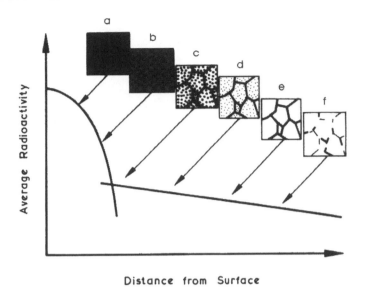

Figure 2.14 Schematic representation of the diffusion profile and tracer distribution for grain boundary diffusion in a polycrystalline compact. The left-hand part of the curve is bell shaped as in Figure 2.2. The insets, (a)–(f), show the distribution of the radioactive tracer in the sample.

Examination of the penetration profile will reveal that it consists of two regions. Initially the curve is bell shaped and dominated by lattice diffusion. At greater penetration, the graph becomes linear. In this region grain boundary diffusion takes over as the major mechanism for the movement of the radioactive tracer atoms.

An experimental penetration profile is shown in Figure 2.15. It refers to the diffusion of the radioactive isotope ^{63}Ni into a polycrystalline block of CoO at 935 °C for 30 min. In order to extract values for the bulk and grain boundary diffusion coefficients from diffusion profiles like that shown in Figure 2.15 it is necessary to solve the appropriate forms of Fick's laws just as we did in section 2.2. The solutions obtained are, not surprisingly, rather complex and will not be given here.

The situation in the case of dislocations is similar to that for grain boundaries. Diffusion down the dislocation core, sometimes called *pipe diffusion*, will be faster than that through the surrounding bulk. If the apparent diffusion coefficient in a material containing many dislocations is compared to the diffusion coefficient in a sample with only a few dislocations a considerable difference will often be found.

The relationship between the *effective* diffusion coefficient, D_E, and the true lattice diffusion coefficient, D_B, is often found to be given by an equation of the form:

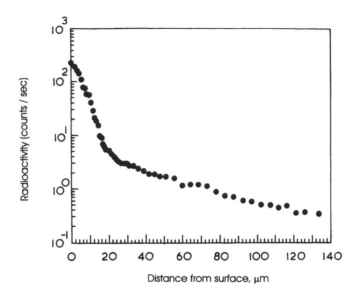

Figure 2.15 Experimental result for the diffusion of radioactive $^{63}Ni^{2+}$ diffusing into polycrystalline CoO at 953 °C. [Redrawn from K. Kowalski, thesis, University of Nancy (1994).]

$$\frac{D_E}{D_B} = 1 + \frac{D_D}{D_B}g$$

where D_D is the diffusion coefficient down the dislocations and grain boundaries and g is a geometrical factor which contains terms for the density of these defects and the number of atom sites taking part in the enhanced diffusion process. Further information on this topic will be found in the supplementary reading section.

2.8   Supplementary reading

There are many books which treat the topic of diffusion in depth. One of the most readable is:

P.G. Shewmon, *Diffusion in Solids*, McGraw–Hill, New York (1963).

The subject is also treated extensively in textbooks of metallurgy and materials science. A good account is given in:

W.D. Kingery, H.K. Bowen and D.R. Uhlmann, *Introduction to Ceramics*, 2nd Edition, Wiley–Interscience, New York (1976).

An advanced review article which summarizes diffusion theory is in:

A.D. LeClaire, in *Treatise on Solid State Chemistry*, ed. N.B. Hannay, Plenum, New York (1976).

Many experimental results are to be found in:

P. Kofstad, *Nonstoichiometry, Diffusion and Electrical Conductivity in Binary Metal Oxides*, Wiley–Interscience, New York (1972).

A clear account of diffusion, together with self-test examples is given by:

R. Metselaar, *J. Mater. Ed.* **6**, 229 (1984); **7**, 653 (1985); **10**, 621 (1988).

Worked examples covering a wide range of diffusion problems is given in the very useful book:

R.G. Faulkener, D.J. Fray and R.D. Jones, *Worked Examples in Mass and Heat Transfer in Materials Engineering*, Institution of Metallurgists, London (n.d.).

Short-circuit diffusion is covered in:

I. Kaur and W. Gust, *Fundamentals of Grain and Interphase Boundary Diffusion*, Ziegler Press, Stuttgart (1988).
A. Atkinson, *J. Chem. Soc. Faraday Trans.* 1307 (1990).

3 The atomic theory of diffusion

3.1 Introduction

In chapter 2 the process of diffusion was presented from an experimental viewpoint. These results are now rationalized on an atomic scale. This has more than an academic interest, for if the atomic movements which constitute diffusion can be understood, then those processes which depend upon diffusion can be controlled more precisely and materials chemically and physically tailored to our special needs.

When diffusion occurs atoms are migrating through the crystal structure. Let us first look at some schematic ways in which we can imagine this to take place. For normal crystals this will be by way of individual atom jumps from one stable position to another.† Some ways in which these individual jumps can take place are illustrated in Figure 3.1.

If Schottky defects are present, as in Figure 3.1(a), atoms or ions can jump from a normal site into a neighbouring vacancy and so gradually move through the crystal. Looking at the pathways shown in Figure 3.1(a), we can see that movement of a diffusing atom into a vacant site corresponds to movement of a vacancy in the other direction. This process is, therefore, frequently referred to as *vacancy diffusion*. In practice it is often very convenient, where vacancy diffusion occurs, to ignore atom movement and to focus attention upon the diffusion of the vacancies as if they were real particles. This mechanism is of importance in close-packed materials.

When Frenkel defects are present, as in Figure 3.1(b), three migration routes are possible. An atom can jump from a normal position into a vacant site created by the Frenkel defect, which is identical to the process of vacancy diffusion just described. Alternatively, an atom in an interstitial site can jump to a neighbouring interstitial position. This is called *interstitial diffusion* and is the mechanism by which tool-steels are hardened. One commonly used process involves heating the steel in an atmosphere of nitrogen. The nitrogen atoms enter the steel and diffuse by an interstitial

---

†In chapter 5 we will focus on fast ion conductors, where cooperative atomic movement can occur involving many atom jumps in concert.

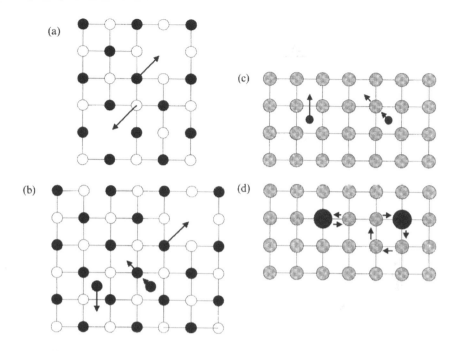

Figure 3.1 Some possible diffusion mechanisms in crystals. (a) Vacancy diffusion in a material containing Schottky defects in an idealized crystal structure. (b) Vacancy, interstitial and interstitialcy diffusion in a material containing Frenkel defects on the cation sub-lattice. (c) Interstitial and interstitialcy diffusion in a material containing interstitial impurities. (d) Ring and exchange diffusion in a material containing substitutional impurities.

mechanism, creating a hard *nitrided* surface layer on the tool. A third mechanism is also possible. Here, an interstitial atom jumps to a filled site and knocks the occupant into an neighbouring interstitial site. This 'knock-on' process is called *interstitialcy diffusion*.

When impurity point defects are present in a crystal they can migrate without the intervention of Schottky or Frenkel defects. Interstitial impurities can move by interstitial and interstitialcy jumps, as shown in Figure 3.1(c). Substitutional impurities can also move by way of two mechanisms, illustrated in Figure 3.1(d). In *exchange diffusion* an impurity swaps places with a neighbouring normal atom while in *ring diffusion* cooperation between several atoms is needed to make the exchange. These processes have been found to take place during the doping of semiconductor crystals.

Remember that in real crystals the paths will be more complex than suggested by the two-dimensional illustrations. It is a good idea to look at a crystal structure model, if you can, to relate the ideas just presented to the three-dimensional geometry of the real world.

3.2   Self-diffusion mechanisms

3.2.1   Energy barriers

The process of self-diffusion involves the movement of atoms in a random fashion through a crystal. No strong driving force, such as a concentration gradient, is present. Nevertheless, each time an atom moves it will have to overcome an energy barrier. This is because the migrating atoms have to leave normally occupied positions which are, by definition, the most stable positions for atoms in the crystal, to pass through less stable positions not normally occupied by atoms. Often atoms may be required to squeeze through a bottle-neck of surrounding atoms in order to move at all.

For a one-dimensional diffusion process we can imagine the energy barrier to take the form shown in Figure 3.2. Referring to this diagram, we can write E_v for the energy barrier to be surmounted by an atom migrating via a vacancy mechanism, E_i for the energy barrier to be surmounted by an interstitial atom and so on. How easily will an atom overcome the barrier it faces? Obviously, the larger the magnitude of E the less chance there is that the atom has the necessary energy to make a successful jump.

We can gain an estimate of this probability by using Maxwell–Boltzmann statistics, which tells us that the probability, p, that a single atom will move from one position of minimum energy in Figure 3.2 to an adjacent position will be given by the equation

$$p = \exp\left(-\frac{E}{kT}\right)$$

where k is Boltzmann's constant and T the absolute temperature. This equation indicates that if E is very small, the probability that the atom will clear the barrier approaches 1.0, if E is equal to kT the probability for a

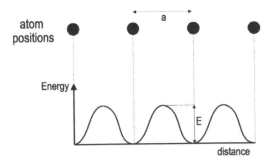

Figure 3.2   Schematic illustration of the potential barrier, E, that a migrating atom has to overcome in moving through a crystal lattice. Stable atom positions, shown by filled circles, are separated by the jump distance, a.

successful jump is about one third and if E increases above kT the probability that the atom could jump the barrier rapidly becomes negligible.

Knowing the probability of an atom clearing the barrier is half of the problem solved. Now it is necessary to know how often the atom tries to make a jump. Recall that the atoms in a crystal are not stationary, but are vibrating continually with a frequency, ν, that is usually taken to have a value of about 10^{13} Hz at room temperature. It is reasonable to suppose that the number of attempts at a jump, sometimes called the *attempt frequency*, will be equal to the frequency with which the atom is vibrating. The number of successful jumps that an atom will make per second, Γ, will be equal to the attempt frequency, ν, multiplied by the probability of a successful move, i.e.

$$\Gamma = \nu \exp\left(\frac{-E}{kT}\right)$$

3.2.2 Atomic migration and diffusion coefficients

In order to link this idea with the magnitude of the self-diffusion coefficient it is necessary to return to the basic equations of diffusion theory, Fick's laws. For the present task the most important of these is *Fick's first law*. For a flow of atoms along the x-direction this can be written as

$$J = -D\frac{dc}{dx}$$

where J is the number of particles crossing a unit area in the solid each second, D is the diffusion coefficient and c is the concentration of the diffusing species at point x after time t has elapsed.

It is not difficult to derive an expression for J for a one-dimensional diffusion process and this is given in Appendix 3.1. The analysis shows that

$$J = -\frac{1}{2}\Gamma a^2 \frac{dc}{dx}$$

Where Γ is the number of successful jumps that an atom makes per second and a is the separation of the stable positions, as in Figure 3.2. If we now compare this equation with Fick's first law, given above, it is clear that:

$$D = \frac{1}{2}\Gamma a^2$$

We have already derived an expression for Γ in terms of the barrier height to be negotiated, E, and substituting, we arrive at

$$D = \frac{1}{2}a^2\nu \exp\left(-\frac{E}{kT}\right)$$

3.2.3   Self-diffusion in crystals

So far our considerations have been limited to diffusion along a single direction. In real crystals, it is necessary to take some account of the three-dimensional nature of the diffusion process. An easy way of doing this is to add a *geometrical factor*, g, into the equation for D so that it becomes

$$D = ga^2\nu \, \exp\left(-\frac{E}{kT}\right)$$

In fact, in the one-dimensional case we have already done this because the factor $1/2$ was a geometrical term to account for the fact that an atom jump can be in one of two directions. In a cubic structure, diffusion can occur along six equivalent directions and a value of g of $1/6$ would be appropriate

$$D = \frac{1}{6}a^2\nu \, \exp\left(-\frac{E}{kT}\right)$$

3.2.4   The effect of the defect population

In a real solid containing a population of defects, the number of jumps per second will not only involve the diffusing species, but also the defect population. Let us consider two examples. If we are discussing interstitial diffusion in AgBr, the amount of diffusion will be decided by the number of Frenkel defects in the system. To take this into account we should incorporate a term for n_f into our equations. Similarly, an atom at a normal lattice position cannot diffuse by a vacancy mechanism unless there is a vacancy population in the crystal. In a pure crystal, this vacancy population will arise from a Schottky defect population, and we would need to include a term for n_s.

The defects of interest could arise in many other ways. If we, therefore, simply express the number of important defects present in the crystal as n, we can correctly write the diffusion coefficient expression as

$$D = ga^2\nu n \, \exp\left(-\frac{E}{kT}\right) \tag{3.1}$$

3.3   The Arrhenius equation and the effect of temperature

If this last equation is compared to the Arrhenius equation

$$D = D_0 \, \exp\left(-\frac{E}{RT}\right)$$

the pre-exponential factor D_0 evaluated experimentally is given by

$$D_0 = ga^2\nu n$$

Similarly, the activation energy, E, is equivalent to the height of the energy barrier to be surmounted. We have, therefore, succeeded in our aim of relating the experimentally observed form of the diffusion equation to an atomic model.

When Arrhenius plots were described in section 2.2, it was remarked that these sometimes fell into two regions; the low temperature part having a rather lower activation energy and the high temperature area having a higher activation energy, as is shown schematically in Figure 3.3. Clearly something is happening at higher temperatures which is using additional energy compared to the low temperature regime. The atomic model allows us to explain this puzzling phenomenon.

We notice in these equations that D_0 includes n, the number of defects present. At low temperatures the number of intrinsic (Frenkel and Schottky) defects will be small. Impurities can also create defects, and it is reasonable to suppose that the defects due to impurities far outnumber the intrinsic defects present. Hence the number of defects, n, in the equation is indeed constant. Thus, in the low temperature part of an Arrhenius plot the activation energy, E, will correspond to height of the energy barrier and D_0 to $ga^2\nu n$.

At high temperatures, however, it is unrealistic to assume that the number of defects, n, is constant. It is more reasonable to assume that at high enough temperatures, the constant n should be replaced by a formula expressing the real population of defects in the crystal. If we take as an example Schottky defects in a crystal of formula MX

$$n_s = N \exp\left(-\frac{\Delta H_s}{2kT}\right)$$

and equation (3.1) becomes

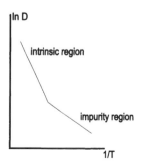

Figure 3.3 Diagram showing the Arrhenius plot expected from a diffusion experiment to find the activation energy for diffusion. The slope in the impurity region yields a value for the energy of movement of the atoms, and the slope in the intrinsic region yields a value for the energy of formation of the defects plus the energy of movement of the atoms.

$$D = g\nu a^2 N \exp\left(-\frac{E_v}{kT}\right)\exp\left(-\frac{\Delta H_s}{2kT}\right)$$

where E_v represents the height of the energy barrier to be overcome in vacancy diffusion. If we have Frenkel defects in a crystal of formula MX

$$n_f = \sqrt{(NN^*)}\exp\left(-\frac{\Delta H_i}{kT}\right)$$

and equation (3.1) becomes

$$D = g\nu a^2 \sqrt{(NN^*)} \exp\left(-\frac{E_i}{kT}\right)\exp\left(-\frac{\Delta H_f}{2kT}\right)$$

where E_i represents the potential barrier to be surmounted by an interstitial atom.

Now both of these equations retain the form

$$D = D_0 \exp\left(-\frac{E}{RT}\right)$$

but now E is the sum of the energy needed to move the defect, E_i or E_v, plus the energy of defect formation. For Frenkel defects

$$E = E_i + \frac{\Delta H_f}{2}$$

and for Schottky defects

$$E = E_v + \frac{\Delta H_s}{2}$$

Thus a comparison of the slope of the Arrhenius plot in the high temperature and the low temperature region will allow an estimate of both the energy barrier and the relevant defect formation energy to be made. Some values found in this way are listed in Table 3.1.

Finally, we should remark on the fact that in our previous discussion we have supposed that the height of the potential barrier will be the same at all temperatures. This is probably not so. As the temperature increases the lattice will expand, and in general E would be expected to decrease. Moreover, some of the other constant terms in the preceding equations will vary slightly with temperature. For example, the lattice spacings will change, leading to a change in the constant a, and the vibration frequency, ν, will increase. The Arrhenius plots reveal this by being slightly curved.

Table 3.1. Some enthalpy values for the formation and movement of vacancies in alkali halide crystals

Material		Schottky defects	
	H_f	H_m cation vacancy	H_m anion vacancy
NaCl	192	84	109
NaBr	163	84	113
KCl	230	75	172
KBr	192	29	46

Material		Frenkel defects	
	H_f	H_m interstitial	H_m vacancy
AgCl	155	13	36
AgBr	117	11	23

All values in $kJ\,mol^{-1}$.

3.4   The relationship between D and diffusion distance

In many processes it is necessary to be able to estimate the distance that atoms can diffuse in a given time. This is important, for example, if we want to know how far a dopant will penetrate into a semiconductor during fabrication. The discussion in chapter 2 revealed that there is no single distance that can be quoted because the distribution of the diffusing atoms has the form of a bell-shaped curve. The development of this curve is shown in Figure 3.4.

In these circumstances, the best we can do is to get an idea of the average distance moved by the diffusing atoms. A straightforward derivation, which assumes a random one-dimensional process is set out in Appendix 3.2. The result obtained, for diffusion along the x-axis only, is:

$$\sqrt{\langle x^2 \rangle} = \sqrt{2Dt}$$

where the square root of $\langle x^2 \rangle$ is a quantity called the *root mean square* value of x, D is the diffusion coefficient and t is the diffusion time. Thus, we find that the root mean square distance that an atom will move is given by a rather simple relationship, and that it is proportional to the square root of the time.

To complete the discussion it is necessary to explain exactly what the root mean square is. To answer this it is necessary to turn to a statistical analysis[†] of the situation. In statistical parlance, the distribution of atoms leading to a bell-shaped curve is called a *normal distribution* or *Gaussian distribution*. The statistics of the normal distribution are well known and tell us that there is a 67% probability that any particular atom will be found in the region

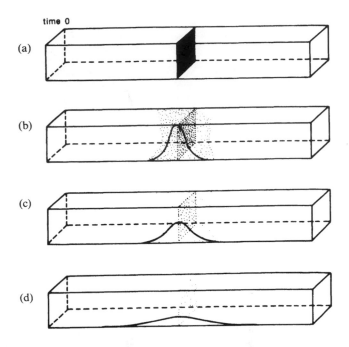

Figure 3.4   (a)–(d) Development of bell-shaped curves by diffusion of a tracer atom in a crystal.

between the starting point of the diffusion and a distance of $\pm\sqrt{\langle x^2 \rangle}$ on either side of it. This region is shown shaded in Figure 3.5. The probability that any particular atom has diffused further than this distance is given by the total area under the curve minus the shaded area, which is 33%. The probability that the atoms have diffused further than $2\sqrt{\langle x^2 \rangle}$ is equal to the total area under the curve minus the area under the curve up to $2\sqrt{\langle x^2 \rangle}$. This is found to be equal to about 95%. Some atoms will have gone further than this distance, but the probability that any one particular atom will have done so is very small.

---

†Interestingly, the equations relevant to this analysis were first derived by Demoivre in 1733, with respect to problems associated with the tossing of coins. There are many similarities, of course. Each atom can jump forwards or backwards, just as a coin can fall heads or tails. A very reasonable model for diffusion can be made simply by placing a row of counters on a board, and tossing a coin to decide if a counter should move forward or backward. After many throws the distribution of the counters will mirror the distribution of atoms.

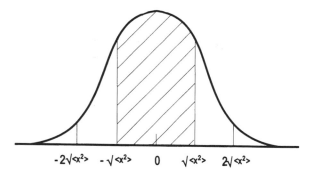

Figure 3.5 A typical bell-shaped normal distribution curve. The shaded region in the centre of the curve indicates a 67% probability that a diffusing atom will be found between the limits $+\sqrt{\langle x^2 \rangle}$ and $-\sqrt{\langle x^2 \rangle}$.

3.5 Correlation effects

So far in this chapter we have discussed the diffusion of atoms in a random fashion throughout the crystal lattice. Each step was unrelated to the one before and not driven by any particular force. The atoms and defects can be considered to be jostled solely by thermal energy. However, diffusion of an atom in a solid is not a truly random process and it is reasonable to suppose that in many circumstances a given jump direction may depend on the direction of the previous jump.

We can explain the situation by considering the vacancy diffusion of an atom in a crystal, as shown schematically in Figure 3.6. The atom we are interested in is dot shaded in Figure 3.6(a), and can be regarded as a tracer atom. It is situated next to a vacant site, so that diffusion can take place. Because we are interested in the diffusion of the tracer, let us assume that it is the tracer that makes the first jump into the vacant site. This leads to the situation shown in Figure 3.6(b). The next jump of the tracer is no longer an entirely random process. It is still next to the vacancy and clearly it is more likely that the tracer will move back to the vacancy, recreating the situation shown in Figure 3.6(a). Hence, of the choices available to the tracer in Figure 3.6(b), a jump back to the situation shown in Figure 3.6(a) is of highest probability.

If we now focus attention upon the vacancy we find a different answer. Considering the situation in Figure 3.6(a), diffusion can occur by way of any of the atoms around the vacancy moving into the empty site. The vacancy, of course, has no preference for any of its neighbours so that its first jump is entirely random. The same is true of the situation shown in Figure 3.6(b). The vacancy will have no need to prefer a jump to the tracer position. Thus, we see that the vacancy can always move to an adjacent cation site, and hence can follow a truly random path.

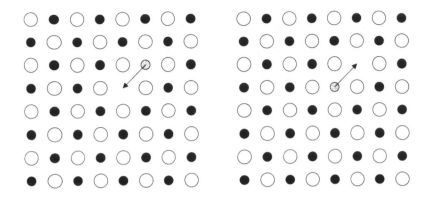

Figure 3.6 Correlated motion in vacancy diffusion. The dot shaded circle represents the tracer atom and the arrows the most likely next jump of the tracer.

When this non-random motion is considered over many jumps, the mean square displacement of the tracer will be less that that of the vacancy, which took the same number of jumps. So it is expected that the observed diffusion coefficient of the tracer will be less than that of the vacancy. In these circumstances, the random-walk diffusion equations need to be modified by the introduction of a *correlation factor, f*. The correlation factor is given by the ratio of the values of the mean square displacement of the tracer to that of the vacancy, provided that the number of jumps considered is large.

$$f = \frac{\langle x^2 \rangle \text{tracer}}{\langle x^2 \rangle \text{vacancy}}$$

Correlation factors for vacancy diffusion generally take values of between 0.5 and 0.8.

If we consider *interstitial diffusion* in which we have only a few diffusing ions and many available empty interstitial sites, we would expect a correlation factor close to 1.0. In effect the interstitial atom moves in a 'sea of vacancies'. In the case of *interstitialcy diffusion*, this will not be true because the number of vacancies will be equal to the number of interstitials present, which will always be rather small in proportion to the number of filled sites.

A number of mathematical procedures have been adopted for evaluating correlation factors. Table 3.2 lists some values for a variety of diffusion mechanisms in some common crystal structure types.

Before concluding this section, it is useful to take the arguments a little further. In our discussion of vacancy diffusion all the cations were assumed to be identical. Often, however, we need to consider the diffusion of an impurity atom in a crystal, say K in NaCl or Ca in MgO. In such cases, the

Table 3.2 Correlation factors for self-diffusion

Mechanism	Structure	Correlation factor (f)
Vacancy	Diamond	0.50
	b.c.c.	0.7272
	f.c.c.	0.7815
	h.c.p.	0.7812 (f_x, f_y)
	h.c.p.	0.7815 (f_z)
Interstitialcy	f.c.c.	0.80
	Fluorite (cations)	1.00
	b.c.c.	0.666
	CsCl (cations)	0.832
	AgBr (cations)	0.666

probability that the impurity will exchange with the vacancy will depend on other factors such as the relative atomic sizes of the impurity compared to the host atoms. In the case of ionic movement, the charge on the diffusing species will play a part. This can be expressed in terms of the jump frequencies of the host and impurity atoms, in which case one is likely to be greater than the other. If, for example, the host atoms have a very high jump frequency, they will be far more likely to move at any instant, giving them a higher correlation factor than the impurities.

3.6 Case history: integrated circuits

Integrated circuits are one of the most important innovations of the twentieth century. They lie at the heart of all computers, smart machines, smart cards and the multiple facets of modern life that rely on information transfer in one way or another. The key step in the manufacture of integrated circuits is diffusion.

Integrated circuit production starts with a slice of single crystal silicon about 125 mm diameter and about 0.2 mm thick. All of the integrated circuitry is fabricated within a p-type or n-type† surface layer of about 6–8 μm thickness on the surface.

This layer is grown on the surface by a process in which volatile compounds of silicon and the chosen impurity are decomposed on the original surface of the single crystal slice. The compounds used are usually halides, which are fairly volatile and can be prepared in a pure state. If the

---

†This terminology will become clear later. For the moment it is only necessary for us to know that conductivity in n-type material is by way of electrons while in p-type material it is by way of electron holes. Conduction by way of holes is analogous to vacancy diffusion.

dopant halide comes from the left of silicon in the periodic table, typically BBr_3, the resulting crystal becomes p-type while if it comes from the right, typically PBr_5, the crystal turns out to be n-type. The growth conditions are carefully chosen so that the single crystal itself grows outward by the addition of new layers of atoms in perfect match with those in the underlying lattice. The dopant is incorporated *substitutionally* on some of the sites normally occupied by the silicon atoms. This process is called *epitaxial growth*. The dopant atoms give the silicon its electrical properties.

Integrated circuits consist of complex patterns of n- and p-type silicon which are grown into the epitaxial layer. These patterns are built up by a repetition of three basic steps. The surface of the slice is oxidized to silicon dioxide, SiO_2. 'Windows' are made in this layer by dissolving away the SiO_2 in precisely specified areas. A dopant is diffused into the exposed silicon to switch its electrical properties from n-type to p-type or vice versa. The role of defects is crucial to each step and the first two steps are both diffusion controlled.

The growth of layers of SiO_2 on the crystal surface is carried out by heating the slice in oxygen gas at about 1200 °C. The growth rate is slow, about 1 μm per h. Now the reason for this can be judged by using the information given in the case study in the previous chapter, mainly concerning Cu_2O. The atoms in SiO_2 are strongly bonded and this material does not contain a high defect population. Furthermore, it is a good electronic insulator and so electronic conduction is not possible. These factors mean that as soon as the initial layer of SiO_2 is formed on the surface of the slice, further transport of Si or O across the layer is very slow indeed. This makes the growth of the layer nicely controllable.

The second key step is the diffusion of the desired n- or p-type dopant into selected exposed regions of silicon. This diffusion step is controlled, as we would expect, by controlling the important diffusion variables of temperature and time. The initial step is to deposit rather a lot of the dopant on the exposed areas of silicon rather rapidly to build up a concentrated surface layer. The next stage is to heat the slice at a controlled temperature to allow the dopant ions to diffuse into the crystal. The equations given for tracer diffusion in the previous chapter will help in working out the important depth–concentration profile which will determine the exact electrical properties of the layer.

Although the fact that the change of electrical behaviour from n-type to p-type and back again is not diffusion controlled, it is an important property. After all, one does not change a metal into an insulator by adding a trace of a non-conducting material or change it back again by adding another trace of metal, yet this is what happens in semiconductors like doped silicon. This is rather an important property which comes about by the nature of the impurity defects and the way in which they control the electrical properties of the material. This unusual state of affairs will be explained in later chapters.

3.7 Ionic conductivity

Ionic movement under the action of an externally applied electric field is called *ionic conductivity*. If two electrodes are introduced into a beaker containing a solution of ions and a voltage is applied, ionic conductivity will occur and the cations will move towards the cathode and the anions towards the anode. This is the origin of the terms cation and anion, in fact. Now the same thing will occur when a voltage is applied across a solid which is made up of ions. In most normal solids, the ionic conductivity is too small to measure at room temperature but all of chapter 5 is devoted to materials in which this limitation has been overcome. Although at first sight it may seem that ionic conductivity in a solid has little connection with self-diffusion, the processes can be treated in very similar ways.

Let us, to illustrate this point, consider diffusion of monovalent ions. The movement will be subject to exactly the same constraints as we discussed previously. The only difference will be that the potential energy barrier E, to be surmounted by the ions in migrating, will be modified in the presence of an applied field V. If the ion is moving against the field, it will have to negotiate not only E but an extra obstacle given by

$$\frac{eaV}{2}$$

where e is the electron charge, a is the normal lattice separation and V is the magnitude of the applied electric field. This situation is shown schematically in Figure 3.7. On the other hand, if movement is in the opposite direction, the applied field helps the ion by lowering the potential barrier by a similar amount. Compared to diffusion, the only new aspect of the situation is that the potential barriers are now tilted in the direction of the applied electric field, as shown in Figure 3.7. At right angles to the field, the energy barrier E remains unchanged. Migration of charged particles in the direction of the field is now favoured.

We can consider the effect this will have upon the movement of the ions by utilizing the same model as we employed to look at self-diffusion. The details are given in Appendix 3.3, the result of which reveals that the ionic conductivity, σ, can be written as

$$\sigma = \left(\frac{\sigma_0}{T}\right) \exp\left(-\frac{E}{kT}\right) \tag{3.2}$$

where E is the height of the diffusion barrier, k is Boltzmann's constant and

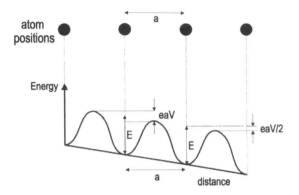

Figure 3.7   Schematic illustration of the potential barrier that a migrating ion must overcome in the presence of an electric field. The values of a, the jump distance, and E, the average height of the potential barrier, are the same as in Figure 3.2, but the effective barrier that an ion faces is lowered for movement in the direction of the field and increased for movement in a direction against the field.

T is the absolute temperature. The term σ_0 is given by

$$\sigma_0 = n\nu a^2 e^2$$

where n is the number of migrating ions per unit volume, ν is the attempt frequency and a and e have already been defined above.

The number of mobile species present in the crystal will depend upon the point defect populations present. At low temperatures, where this population is controlled by the impurities present, n will be constant. This means that we can use an Arrhenius-like plot to determine a value for E in the following way. Multiplying both sides by T and taking logarithms in equation (3.2) gives

$$\ln(\sigma T) = \ln(\sigma_0) + \frac{-E}{kT}$$

Hence a plot of $\ln(\sigma T)$ versus $1/T$ will have a slope equal to $-E/k$.

At high temperatures it is reasonable to suppose that the value of n will increase due to the creation of extra Frenkel or Schottky defects. In this case we can substitute for n from the equations given in chapter 1, as we did above.

For Frenkel defects

$$E = E_i + \frac{\Delta H_f}{2}$$

and for Schottky defects

$$E = E_v + \frac{\Delta H_s}{2}$$

If both high and low temperature regimes are important, the Arrhenius plots for such materials will show a knee similar to that shown in Figure 3.4.

Thus one is able to measure values for enthalpy of formation and migration of point defects using conductivity data in an analogous way to using diffusion data. In fact, although conductivity measurements are not always easy to make, they are usually much easier to perform than diffusion experiments, and are usually preferred. To illustrate this, we reproduce in Figure 3.8 some experimental results for the ionic conductivity of NaCl which show the intrinsic and impurity regions clearly.

The equations that have been derived in this section are correct for monovalent ions only. In order to make them applicable to ions of formal charge $+z$, we must replace the term e by $+ze$.

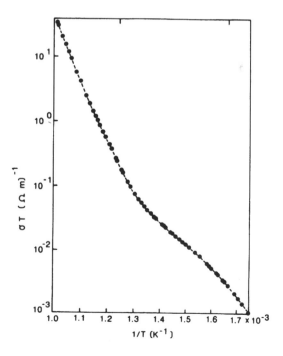

Figure 3.8 The ionic conductivity of NaCl clearly showing the intrinsic and impurity regions. [From R. Kirk and P.L. Pratt, *Proc. Br. Ceram. Soc.* **9**, 215 (1967).]

3.7.1 The relationship between ionic conductivity and diffusion coefficient

From our foregoing discussions, ionic conductivity and ionic diffusion are seen to be closely related. If both processes occur by the same random-walk mechanism, the relationship between the self-diffusion coefficient, D, and the ionic conductivity, σ, can readily be derived.

The ionic conductivity of a monovalent ion in a direction parallel to an external electric field is given by

$$\sigma = \left(\frac{n\nu a^2 e^2}{kT}\right)\exp\left(-\frac{E}{kT}\right)$$

The equivalent equation for diffusion of an ion, moving in one direction, over an identical potential barrier is given by

$$D = \nu a^2 \exp\left(\frac{-E}{kT}\right)$$

Combining these two equation gives

$$\frac{\sigma}{D} = \frac{ne^2}{kT}$$

This equation, first derived by Einstein, is a simplified form of an equation generally known as the *Nernst–Einstein* equation. For an ion of charge $+z$, the equation becomes

$$\frac{\sigma}{D} = \frac{nz^2 e^2}{kT}$$

In both these equations n is the number of mobile ions of charge ze per unit volume.

Example 3.1

Estimate the value of σ/D.

Taking the charge on the electron to be 1.6×10^{-19} C and Boltzmann's constant to be 1.38×10^{-23} J K^{-1}

$$\frac{\sigma}{D} = \frac{n \times \left(1.6 \times 10^{-19}\right)^2}{(1.38 \times 10^{-23})T}$$

$$= 1.86 \times 10^{-15}\frac{n}{T}$$

In general, n takes a value of approximately 10^{22} defects m^{-3}, and taking T as 1000 K

$$\frac{\sigma}{D} = 1.68 \times 10^4 \, \text{C}^2 \, \text{J}^{-1} \, \text{m}^{-3}$$

Hence we see that D is considerably smaller than σ, so not only are conductivity values more easily obtained experimentally, but are also a lot larger in magnitude.

3.7.2 Transport numbers

The conventional method of stating the electrical conductivity of a crystal gives no indication of the component conductivities which may be contributing to the overall effect. Conductivity could arise from either cations, anions or electrons. *Transport numbers* give the extent to which each of these factors contribute to the conductivity. Thus, if we write the total conductivity of a material as σ we can write

$$\sigma = \sigma_{\text{cation}} + \sigma_{\text{anion}} + \sigma_{\text{electron}}$$

$$\sigma_{\text{cation}} = t_{\text{cation}} \sigma$$

$$\sigma_{\text{anion}} = t_{\text{anion}} \sigma$$

$$\sigma_{\text{electron}} = t_{\text{electron}} \sigma$$

where σ_{cation}, σ_{anion} and σ_{electron} are the conductivities of the cations, anions and electrons, and t_{cation}, t_{anion} and t_{electron} are called the transport numbers for cations, anions and electrons, respectively. As can be seen from these relationships

$$\sigma = \sigma(t_{\text{cation}} + t_{\text{anion}} + t_{\text{electron}})$$

$$t_{\text{cation}} + t_{\text{anion}} + t_{\text{electron}} = 1$$

Among the compounds of interest in this book the following generalizations hold.

Halides rarely show electronic conductivity, so that t_{electron} is 0. Li halides in which the small Li^+ ion is very mobile, and Ag halides with Frenkel defects on the cation sub-lattice, have $t_{\text{cation}} = 1.0$. Ba and Pb halides, with very large cations and which contain Frenkel defects on the anion sub-lattice, show only anion migration and hence have $t_{\text{anion}} = 1.0$. NaF, NaCl, NaBr and KCl in which Schottky defects prevail and in which the cations and anions are of similar sizes, have both cation and anion contributions to ionic conductivity.

Oxides and sulphides, especially of transition metals, often have appreciable values of $t_{electron}$.

3.8   Supplementary reading

Atomic diffusion is covered in many textbooks of materials science and physical chemistry. For a very clear introductory article, see:

Phase Diagrams and Microstructure, Open University Introduction to Materials course, Unit 5, Open University (1974).

A more comprehensive discussion is included in:

W.D. Kingery, H.K. Bowen and D.R. Uhlmann, *Introduction to Ceramics*, 2nd Edition, Wiley–Interscience, New York (1976).

At an advanced, but still readable level, refer to:

P.G. Shewman, *Diffusion in Solids*, McGraw-Hill (1963).
J.R. Manning, *Diffusion Kinetics for Atoms in Crystals*, Van Nostrand (1968).

A clear exposition of the principles of diffusion, together with self-assessment questions, is given by:

R. Metselaar, *J. Mater. Edn.* **6**, 229 (1984); **7**, 653 (1985); **10**, 621 (1988).

The supplementary reading at the end of chapter 2 also contains material relevant to the present chapter.

Appendix 3.1
Atomic migration and the diffusion coefficient

We can derive expressions for J and dc/dx by reference to Figure A1. Here we have adjacent planes in a crystal, numbered 1 and 2, separated by the atomic jump distance, a. Let n_1 and n_2 be the numbers of diffusing atoms per

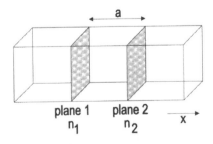

Figure A1   Schematic diagram of two adjacent planes in a crystal, 1 and 2, separated by the jump distance for diffusion, a. The number of diffusing atoms on these planes is n_1 and n_2 per unit area, respectively.

unit area in planes 1 and 2, respectively. If Γ_{12} is the frequency with which an atom moves from plane 1 to plane 2, then the numbers of atoms moving from plane 1 to 2 per second is j_{12}, where

$$j_{12} = n_1\Gamma_{12}$$

Similarly, the number moving from plane 2 to plane 1 is j_{21} where

$$j_{21} = n_2\Gamma_{21}$$

The net movement, often called the *flux*, between the planes, J, is given by

$$J = j_{12} - j_{21} = (n_1\Gamma_{12} - n_2\Gamma_{21})$$

If the process is random, the jump frequency is independent of direction and we can set Γ_{12} equal to Γ_{21}. Moreover, if the jump frequency is independent of direction, then half of the jumps, on average, will be in one direction and half will be in the opposite direction, so we can write

$$\Gamma_{12} = \Gamma_{21} = \frac{1}{2}\Gamma$$

where Γ represents the overall jump frequency of the diffusion atoms, i.e.

$$J = \frac{1}{2}(n_1 - n_2)\Gamma$$

To proceed further, we must relate n_1 and n_2 to the concentration of mobile atoms in the crystal. This is readily accomplished. The number of mobile atoms on plane 1 is n_1 per unit area, so that the concentration per unit volume at plane 1 is n_1/a, which we will call c_1. Similarly, the number of mobile atoms per unit area on plane 2 is n_2, so that the concentration per unit volume at plane 2 is n_2/a which we will call c_2. Thus

$$(n_1 - n_2) = a(c_1 - c_2)$$

Hence

$$J = \frac{1}{2}a(c_1 - c_2)\Gamma$$

The concentration gradient, dc/dx, is given by the change in concentration between planes 1 and 2 divided by the distance between planes 1 and 2, that is

$$-\frac{dc}{dx} = \frac{(c_1 - c_2)}{a}$$

where a minus sign is introduced as the concentration falls as we move from plane 1 to plane 2. Hence

$$(c_1 - c_2) = -a\frac{dc}{dx}$$

and

$$J = -\frac{1}{2}\Gamma a^2 \frac{dc}{dx}$$

If we now compare this equation with Fick's first law

$$J = -D\frac{dc}{dx}$$

it is clear that

$$D = \frac{1}{2}\Gamma a^2$$

We have already derived an expression for the jump frequency, Γ, in terms of the barrier height to be negotiated, E, and so

$$D = \frac{1}{2}a^2\nu\left(-\frac{E}{kT}\right)$$

Appendix 3.2
The relationship between D and diffusion distance

Suppose an atom is moving from one stable site to the next in the x-direction by way of a random walk. The net displacement of a diffusing atom after N jumps will be the algebraic sum of the individual jumps. If x_i is the distance moved along the x axis in the ith jump, the distance moved after a total of N jumps, x will simply be the sum of all the individual steps, i.e.

$$x = x_1 + x_2 + x_3... = \sum x_i$$

In our case each individual value of x_i can be $+a$ or $-a$.

If the jumps take place with an equal probability in both directions, after N jumps the total displacement may have any value between zero and Na. In order to proceed, we use a mathematical short cut. If the jump distances are squared, we automatically get rid of all the negative quantities. So, squaring the summation

$$\begin{aligned}
x^2 =& (x_1 + x_2 + x_3...x_N)(x_1 + x_2 + x_3...x_N) \\
=& (x_1x_1 + x_1x_2 + x_1x_3...x_1x_N \\
& + x_2x_1 + x_2x_2 + x_2x_3...x_2x_N \\
& + ... \\
& + x_Nx_1 + x_Nx_2 + x_Nx_3...x_Nx_N)
\end{aligned}$$

We can write this in a more condensed form as

$$x^2 = \sum x_i^2 + 2\sum x_i x_{i+1} + 2\sum x_i x_{i+2} + \dots$$

$$= \sum x_i^2 + 2\sum\sum x_i x_{i+j}$$

If a large number of jumps is assumed, and knowing that each jump may be either positive or negative, the double sum terms in the last equation average to zero. The result is called the *mean square displacement*, and written $\langle x^2 \rangle$. The equation therefore reduces to the manageable form

$$\langle x^2 \rangle = \sum x_i^2$$

As each jump, x_i, can be equal to $+a$ or $-a$,

$$\langle x^2 \rangle = x_1^2 + x_2^2 + x_3^2 \dots + x_N^2$$

$$= a^2 + a^2 + a^2 \dots + a^2$$

i.e. $\langle x^2 \rangle = Na^2$

In section 3.2 we defined Γ as the frequency with which an atom jumps from one site to another along the x-direction, so that the total number of jumps, N, will be given by Γ jumps per second multiplied by the time, t, over which the diffusion experiment has lasted, that is

$$N = \Gamma t$$

Hence

$$\langle x^2 \rangle = \Gamma t a^2$$

However, we determined that the term Γa^2 is equal to $2D$, so that

$$\sqrt{\langle x^2 \rangle} = 2Dt$$

The average distance that an atom will travel in time t is the square root of $\langle x^2 \rangle$, a quantity called the *root mean square* value of x, which is given by

$$\sqrt{\langle x^2 \rangle} = \sqrt{2Dt}$$

Appendix 3.3
Ionic conductivity

The ionic conductivity, σ, can be defined in terms of the equation

$$\sigma = ne\mu$$

where n is the number of migrating monovalent ions per unit volume, each carrying a charge e and μ is called the *mobility* of the ion. The strategy is to use the idea of atomic jumps to estimate the mobility of the ions and hence the ionic conductivity.

The number of jumps that an ion will make in the direction of the field per second is given by a modified form of the equation

$$\Gamma = \nu \exp\left(-\frac{E}{kT}\right)$$

For movement in the favoured direction, the number of successful jumps that an ion will make will be

$$\Gamma_+ = \nu \exp\left[-\frac{\left(E + \frac{1}{2}eaV\right)}{kT}\right]$$

where we have simply substituted the new potential barrier, $E - \frac{1}{2}eaV$ for E. In a direction against the field the number of successful jumps will be given by

$$\Gamma- = \nu \exp\left[-\frac{\left(E + \frac{1}{2}eaV\right)}{kT}\right]$$

where we have a similar substitution for E but this time of $E + \frac{1}{2}eaV$, as the barrier height is now increased.

The overall jump rate in the direction of the field is $\Gamma_+ - \Gamma_-$, and as the net velocity of the ions in the direction of the field, *vel*, is given by the net jump rate multiplied by the distance moved at each jump we can write

$$vel = \nu a \left\{ \exp\left[-\frac{\left(E - \frac{1}{2}eaV\right)}{kT}\right] - \exp\left[-\frac{\left(E + \frac{1}{2}eaV\right)}{kT}\right] \right\}$$

$$= \nu a \exp\left(-\frac{E}{kT}\right)\left[\exp\left(\frac{eaE}{2KT}\right) - \exp\left(-\frac{eaE}{2kT}\right)\right]$$

For low field strengths eaV is much less than kT, and $\exp(eaV/2kT) - \exp(-eaV/2KT)$ may be replaced by (eaV/kT)† as we can see from Table A1. We can now proceed to write

$$vel = \left(\frac{\nu a^2 eV}{kT}\right)\exp\left(-\frac{E}{kT}\right)$$

---

†To see just what sort of field strength this approximation corresponds to, we can take a value of eaV/kT equal to 1 to be the maximum value at which the approximation holds and estimate the corresponding field strength. Taking a temperature of 500 K, and a value of a of about 0.3×10^{-9} m yields a value for V of 2.87×10^2 V cm^{-1}. Thus, the approximation is a reasonable one for the field strengths up to about 300 V cm^{-1}.

Table A1 The equivalence of eaE/kT and $\left[\exp\left(\dfrac{eaE}{2kT}\right) - \exp\left(\dfrac{-eaE}{2kT}\right)\right]$

$\dfrac{eaE}{kT}$	$\exp\left(\dfrac{eaE}{2kT}\right)$	$\exp\left(\dfrac{-eaE}{2kT}\right)$	Difference
0.001	1.00100	0.99900	0.00200
0.01	1.01005	0.99005	0.02000
0.1	1.10517	0.90484	0.20033
1.00	2.71828	0.36788	2.35040
10.00	22026.47	4.54×10^{-5}	22026.47

The *mobility*, μ, of the ion is defined as the rate of movement when the value of V is unity, so

$$\mu = \left(\frac{va^2e}{kT}\right)\exp\left(-\frac{E}{kT}\right)$$

Returning to the equation for the ionic conductivity, σ, given by

$$\sigma = ne\mu$$

and substituting for μ, we can write

$$\sigma = \left(\frac{nva^2e^2}{kT}\right)\exp\left(-\frac{E}{kT}\right)$$

We can see that this equation takes on a form

$$\sigma = \left(\frac{\sigma_0}{T}\right)\exp\left(-\frac{E}{kT}\right)$$

In this equation, σ_0 includes a term, n, which is the number of mobile species present in the crystal. At low temperatures n will be controlled by the impurity population. When atom migration takes place via a vacancy diffusion mechanism we can write

$$\sigma_v = \left[\frac{n_v va^2e^2}{kT}\right]\exp\left(-\frac{E_v}{kT}\right)$$

and when it takes place by the migration of interstitials

$$\sigma_i = \left[\frac{n_i va^2e^2}{kT}\right]\exp\left(-\frac{E_i}{kT}\right)$$

In this regime we can obtain measures of E_v or E_i directly from the Arrhenius-like plots of $\ln \sigma T$ versus $1/T$. At high temperatures it is reasonable to suppose that the values of n_v or n_i are temperature dependent.

In this case, we can substitute for n from the equations given in chapter 1 to obtain, for interstitials due to Frenkel defects

$$\sigma_i = \left[\frac{\nu a^2 e^2 (NN^*)^{1/2}}{kT} \right] \exp\left(-\frac{E_i}{kT} \right) \exp\left(-\frac{\Delta H_f}{2kT} \right)$$

and for vacancies due to Schottky defects

$$\sigma_v = \left[\frac{\nu a^2 e^2 N^2}{kT} \right] \exp\left(-\frac{E_v}{kT} \right) \exp\left(-\frac{\Delta H_s}{2kT} \right)$$

In this case, an Arrhenius plot of $\ln \sigma T$ versus $1/T$ will yield a higher value for the activation energy than in the low temperature region. The new value for E will be composed of two terms

$$E_s = E_v + \frac{\Delta H_s}{2}$$

for Schottky defects and

$$E_f = E_i + \frac{\Delta H_f}{2}$$

for Frenkel defects.

4   Non-stoichiometry and point defects

4.1   The composition of solids

Classical chemistry has taught people to think in terms of compounds in which the ratios of the atomic components are small integers. The compositions of such compounds are, moreover, thought to be totally fixed. Examples include molecules such as HCl, H_2O and NH_3. Experimentally it has been found that this does not apply very well to numerous solid phases. Phases with a variable composition are called *non-stoichiometric compounds*. When it is necessary to stress the non-stoichiometric nature of a material the symbol $\sim$ will prefix the formula. For example, $\sim$FeO refers to non-stoichiometric iron monoxide which is only approximately represented by the formula FeO. In the rest of this book, the underlying reasons for non-stoichiometry are described. The approach taken is to describe non-stoichiometric materials as normal crystals which contain a varying population of defects. As we will see, these defects endow the materials with important and often very surprising properties indeed.

4.2   Solid solutions

One of the simplest ways to vary the composition of a solid is to replace one ion by another of the same size and charge. We have looked at this from the point of diffusion in chapter 2, when we considered what would happen if a crystal of MgO was placed in contact with a crystal of NiO and heated. Looking at matters from a structural perspective, addition of a little NiO to MgO followed by heating will yield an X-ray powder pattern which is virtually identical to that of pure MgO. This will continue all across the range from pure MgO to pure NiO. A *solid solution* is said to have formed. In this case both MgO and NiO have the same *rock salt* structure and at intermediate compositions the product is a material with the *rock salt* structure but with a mixture of Mg^{2+} and Ni^{2+} distributed over the cation sites in the proportions given by the starting composition before heating. The formation reaction can be written

$$(1 - x)\text{MgO} + x\text{NiO} \longrightarrow \text{Mg}_{1-x}\text{Ni}_x\text{O}$$

The phase formed can be given the formula $Mg_{1-x}Ni_xO$ where x can take values of between zero and 1.

The solid solution is shown schematically in Figure 4.1. In this case it has been assumed that the crystal is fairly pure MgO. The Ni^{2+} cations present replace some of the Mg^{2+} cations and so are *substitutional impurity defects* of the same general sort as discussed in chapter 1. The impurity cations give the crystals new properties, in this case the most obvious of these is colour, as the mixed crystals take on a green hue.

Solid solutions have been widely studied because they provide a convenient means of changing the physical properties of a material in a controlled way. Interestingly, the properties of the resultant solid solutions are not just a weighted average of the properties of the phases which occur at each end of the solid solution range, and surprising new features are often found. This effect will be illustrated in a seemingly very simple system, that between Al_2O_3 and Cr_2O_3 in chapter 7, where the reason for the formation of rubies and ruby lasers is explained. Another important series of solid solutions with variable magnetic properties is described in the case study following this section.

4.2.1 Vegard's law

If the unit cell size of a solid solution is measured as a function of the composition, it will frequently be found to obey *Vegard's law*. The 'ideal law', which was first propounded in 1921, states that the lattice parameter of a solid solution of two phases with similar structures will be linear function of the lattice parameters of the two end members of the composition range. This is illustrated in Figure 4.2 for the solid solution which forms between the two *rock salt* structure oxides CoO and MnO. Even when Vegard's law

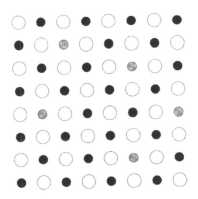

Figure 4.1 A schematic illustration of a solid solution $Mg_{1-x}Ni_xO$. The O^{2-} ions are drawn as large open circles, the Mg^{2+} ions as filled circles and the Ni^{2+} ions as shaded circles.

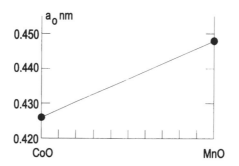

Figure 4.2 The Vegard's law dependence of the lattice parameter of the cubic *rock salt* solid solution between CoO (a = 0.4260 nm) and MnO (a = 0.44448 nm).

in its ideal form is not obeyed exactly, the line joining the lattice parameters of the parent phases is often a shallow curve. In either case, Vegard's law can be used to determine the composition of intermediate compositions in a solid solution quite easily. It is a straightforward matter to obtain an X-ray powder pattern and determine the unit cell dimensions of a solid solution. A comparison of the result with the unit cell dimensions of the parent compounds will give a reliable value for the relative percentages of the components.

Example 4.1

The unit cell size of CaO is 0.48105 nm and that of SrO is 0.51602 nm. Both adopt the *rock salt* type structure. Estimate the composition of a crystal of formula $Ca_xSr_{1-x}O$ which was found to have a unit cell of 0.5003 nm.

   The easiest way to tackle this is to draw a straight-line graph connecting the two lattice parameters, as shown in Figure 4.2, and read off the composition along the x-axis. The formula of the crystal is found to be $Ca_{0.65}Sr_{0.45}O$.

4.3   Case study: magnetic spinels

Solid solutions of a family of oxides called *spinels* with a general formula AB_2O_4 are of considerable commercial importance. Among the most useful are those which contain iron as one cation in the solid solution, as they show a range of useful magnetic properties. Indeed, the first magnetic material known, and for many centuries the only magnetic material known, was the magnetic oxide Fe_3O_4 which is also called *lodestone* or *magnetite*. The name lodestone is a corruption of the expression *leading stone*, which reflects on the fact that one of its earliest and most important uses was as a compass

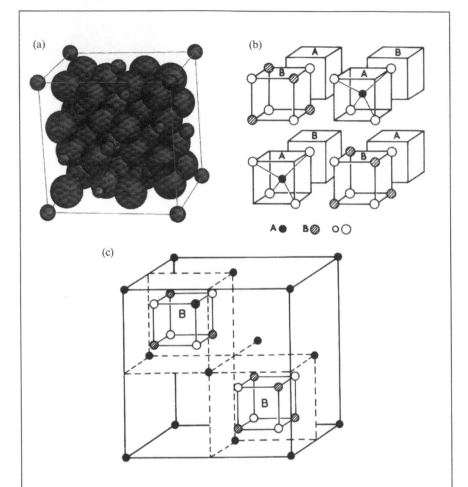

Figure 4.3 (a) A perspective view of a unit cell of *normal* spinel structure $A[B_2]O_4$. The A^{2+} cations occupy tetrahedral sites and the B^{3+} cations occupy octahedral sites in the oxygen array. The structure is built from octants of AO_4 tetrahedra and B_4O_4 cubes as shown in (b). These are arranged in a face-centred cubic array of A^{2+} cations as shown in (c).

The spinel structure

The *spinel* family of oxides with composition AB_2O_4 takes its name from the mineral spinel, $MgAl_2O_4$. The unit cell is cubic with a lattice parameter of 0.809 nm. Each unit cell contains eight $MgAl_2O_4$ formula units, that is, eight Mg^{2+} ions, 16 Al^{3+} ions and 32 O^{2-} ions. A wide variety of cations can adopt the *spinel* structure, but the largest group is formed by a combination of A^{2+} and B^{3+} cations. All crystallize with the structure shown in Figure 4.3. In this structure, some of the cations are surrounded by four oxide ions in what are called *tetrahedral sites* while twice as many are surrounded by six oxide ions in *octahedral sites*. The oxygen anions are in the same arrangement as in the *rock salt* structure and this geometry leads to one octahedral site and two tetrahedral sites available per anion. The spinel unit cell contains four AB_2O_4 formula units, which means that there are 32 oxygen ions present and so 32 octahedral sites and 64 tetrahedral sites are available. In the structure, one-eighth of the tetrahedral sites and one-half of the octahedral sites are filled by the cations.

There are a number of ways in which this can be done. If all of the A^{2+} cations are located in the tetrahedral sites and the B^{3+} ions fill the octahedral sites we have the *normal spinel structure*, written as $A^{2+}[B_2^{3+}]O_4$ where the square brackets enclose the octahedrally co-ordinated ions. If the A^{2+} cations occupy most of the octahedral sites, and the B^{3+} cations fill up the rest of the octahedral sites and all of the tetrahedral sites, the arrangement is called the *inverse spinel* structure, written as $B^{3+}[A^{2+}B^{3+}]O_4$. In $MgAl_2O_4$, the situation is not quite so clear cut as some Mg ions do occupy octahedral sites, while most sit in the tetrahedral sites. This intermediate situation can be described via an occupation factor, λ, which gives the fraction of B^{3+} cations in the tetrahedral sites. A normal spinel has a value of 0 for λ, while an inverse spinel has a value of 0.5. $MgAl_2O_4$ has a λ value of about 0.05

pointer. Small fragments of lodestone floating in a basin of water always came to rest pointing north–south. This property was much prized by early navigators who made use of the material as a primitive compass. More recently, large tonnages of iron *spinels* are used as transformer cores, deflection yoke cores and beam-focusing coils in television sets. The recording heads on tape and video-recorders are made of the same group of materials. In order to understand how these useful properties come about, it is necessary to examine the structures of these solid solutions in detail.

The magnetically important *ferrites* have a general formula $A^{2+}Fe_2^{3+}O_4$. Many crystallize with the inverse spinel structure and can be written as $Fe^{3+}[A^{2+}Fe^{3+}]O_4$. In these materials the magnetic moment of the Fe^{3+} ions in the tetrahedral sites is opposed to that of the Fe^{3+} ions in the octahedral sites, so that the net magnetic moment due to Fe^{3+} is zero. Take, for example, nickel ferrite. This can be written as $Fe^{3+}\uparrow[Fe^{3+}\downarrow Ni^{2+}\uparrow]O_4$

and the net magnetic moment is due to the Ni^{2+} ions alone. Other ferrites with the inverse structure are formed by Mn^{2+} and Co^{2+}. The oxide Fe_3O_4 is an inverse spinel, although this fact is obscured in the formula by the fact that the A and B cations are both Fe. The correct state of affairs is given when the formula is written $Fe^{3+}\uparrow[Fe^{3+}\downarrow Fe^{2+}\downarrow]O_4$. Note that the overall magnetic moment of the compound is due to the Fe^{2+} contribution, as with $NiFe_2O_4$. The magnetic behaviour exhibited by this type of ordered arrangement of spins is called *ferrimagnetism*.

The object of a large amount of research on these compounds centres on how to make solid solutions which will modify this magnetic situation to produce compounds with a precise and unique magnetic signature. In effect, the strategy is to introduce substitutional impurities to achieve this objective. The structure gives us four degrees of freedom to explore, magnetic or non-magnetic impurity cations on tetrahedral sites, and magnetic or non-magnetic impurities on octahedral sites. Because the magnetic moments on all of these cations is different, it is possible to make solid solutions with quite precise magnetic behaviour. In effect, we are able to introduce controlled amounts of magnetic defects (or magnetism) into the solid solution in this way.

The flexibility of the system can be increased greatly when solid solutions are made with normal ferrites. An example is provided by the normal *spinel* $ZnFe_2O_4$. In this material, the magnetic moments of the individual Fe^{3+} ions are opposed even though they both occupy the octahedral sites, and the formula can be written $Zn^{2+}[Fe^{3+}\uparrow Fe^{3+}\downarrow]O_4$. Suppose that this material is now reacted to form a solid solution with $NiFe_2O_4$ thus

$$(1-x)NiFe_2O_4 + xZnFe_2O_4 \longrightarrow Zn_xNi_{1-x}Fe_2O_4$$

The rather straightforward formula hides an intriguing situation. This is made clear if we write out the formula as

$$Zn_x^{2+}Fe_{1-x}^{3+}\uparrow\left[Fe^{3+}\downarrow Ni_{1-x}^{2+}\uparrow Fe_x^{3+}\uparrow\right]O_4$$

where the arrows represent the overall magnetic moment of the ions. When x is zero the material has a magnetism due only to the Ni^{2+} ions. When x is 1.0 the material is non-magnetic. In between, the magnetism is a steady function of the Ni^{2+} concentration and Ni^{2+} is the *magnetic defect*. But what happens if the formula is

$$Zn_x^{2+}Fe_{1-x}^{3+}\uparrow\left[Fe^{3+}\uparrow Ni_{1-x}^{2+}\uparrow Fe_x^{3+}\downarrow\right]O_4$$

Now when x is zero all the magnetic moments point in the same direction and we have a much higher magnetism than in $NiFe_2O_4$. Experimentally, it is found that the defect interactions are complex and the observed magnetism rises to a maximum when x is near to 0.5. Nickel–zinc ferrites of about this composition are used in recording heads for audio and video-recorders.

4.4 Non-stoichiometry

Solid solution formation involves the substitution of one cation or anion by another of the same size and charge. What would happen if the charges on the substituted ions did not match? In such cases the charge imbalance is offset by a change in composition. Materials which have a composition range are called *non-stoichiometric compounds*. The variation in composition need not be by substitution only. Other mechanisms, such as the addition of interstitial ions, can also lead to the formation of non-stoichiometric compounds. Thus, the appearance of non-stoichiometry will be closely associated with the presence of substitutional or interstitial defects in the structure.

Non-stoichiometric compounds are important because the composition variation endows the compounds with useful and often fascinating properties. Moreover, these can be modified by varying the relative proportions of the atomic constituents and in controlling the types of defects present. In this volume, we intend to focus most attention on inorganic non-metallic materials and so a short but by no means exhaustive list of some of these particular non-stoichiometric phases is given in Table 4.1. The interacting role of defects and composition in non-stoichiometric compounds will be explained in the following sections.

4.5 Substitutional impurities

4.5.1 Vacancy formation

Think about the situation that occurs when a cation of higher valence is introduced into a compound. For example, suppose a crystal of NaCl contains a small amount of Ca^{2+} as a substitutional impurity replacing some Na^+ ions present in the sample. In this case, each Ca^{2+} ion in the crystal will increase the total amount of positive charge present. Now common observation will reveal that such crystals do not show an excess of positive charge proportional to the impurity Ca^{2+} concentration and a compensating mechanism must operate to restore neutrality. One way to achieve this balance this is to create one vacancy among the Na^+ cations for each Ca^{2+} incorporated into the structure, as shown in Figure 4.4. Thus, it appears that the substitution has produced a double defect population, substitutional impurities and an equal number of cation vacancies. In fact, this is found rather often and a general rule is that if we dope with a cation

Table 4.1 Approximate composition ranges for some non-stoichiometric phases†

Compound	Approximate formula	Composition range
Rock salt structure oxides		
TiO_x	TiO	$0.65 < x < 1.25$
VO_x	VO	$0.79 < x < 1.29$
Mn_xO	MnO	$0.848 < x < 1.00$
Fe_xO	FeO	$0.833 < x < 0.957$
Co_xO	CoO	$0.988 < x < 1.000$
Ni_xO	NiO	$0.999 < x < 1.000$
Fluorite structure oxides		
CeO_x	Ce_2O_3	$1.50 < x < 1.52$
	$Ce_{32}O_{58}$	$1.805 < x < 1.812$
ZrO_x	ZrO_2	$1.700 < x < 2.004$
UO_x	UO_2	$1.65 < x < 2.25$
Oxide 'bronzes'		
$Li_xV_2O_5$		$0.2 < x < 0.33$
$Na_xV_2O_5$		$0.13 < x < 0.31$
$Li_xV_3O_8$		$1.13 < x < 1.33$
Li_xWO_3		$0 < x < 0.50$
Ca_xWO_3		$0 < x < 0.125$
In_xWO_3		$0.20 < x < 0.33$
Sulphides		
TiS_x	TiS	$0.971 < x < 1.064$
	Ti_8S_9	$1.112 < x < 1.205$
	Ti_3S_4	$1.282 < x < 1.300$
	Ti_2S_3	$1.370 < x < 1.587$
	TiS_2	$1.818 < x < 1.923$
Nb_xS	NbS_2	$0.92 < x < 1.00$
Ta_xS_2	TaS_2	$1.00 < x < 1.35$
$Ba_3Fe_{1+x}S_5$	$Ba_6Fe_3S_{10}$	$0 < x < 1$

†Note that all composition ranges are temperature dependent and the figures given here are only intended as a guide.

of higher valence than the normal cations present, a vacancy population on the cation sub-lattice is also produced. This technique has great potential as a means of modification of the properties of crystals, as will be seen later.

4.5.2 Calcia-stabilized zirconia

Calcia-stabilized zirconia provides a good example of the consequences of incorporating an ion with a lower valence into a crystal structure. Zirconia, ZrO_2, is an important material as it remains inert and stable even at temperatures of up to 2500 °C. This means that it finds uses in a diverse range of applications from rocket motors to furnace linings. The difficulty with pure zirconia, however, is that it will not tolerate cycling from high to low temperatures repeatedly. This is because the structure at room

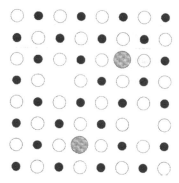

Figure 4.4 A schematic illustration of vacancy formation in NaCl doped with Ca^{2+} ions. The Na^+ ions are shown as filled circles, the Cl^- ions as open circles and the Ca^{2+} impurities as large shaded circles.

temperature has a monoclinic unit cell which changes to tetragonal above about 1000 °C. This change is accompanied by a change in the unit cell volume. The resulting stress caused by the transition causes the material to fracture.

The solution to the problem is to react the zirconia with calcia, CaO. The product, which has a cubic unit cell, is called *calcia-stabilized zirconia*. X-ray diffraction shows this to be a non-stoichiometric phase which exists between the approximate limits of 16 and 28 mole % CaO. The unit cell dimensions of this cubic phase change smoothly across the composition range in agreement with Vegard's law. Because the cubic symmetry does not change with temperature the material can be used in high temperature applications without a problem.

Cubic calcia-stabilized zirconia crystallizes with the *fluorite* structure, which is shown in Figure 4.5. It is found that the Ca^{2+} cations occupy positions that are normally filled by Zr^{4+} cations, that is, cation substitution has occurred. As the Ca^{2+} ions have a lower charge than the Zr^{4+} ions the crystal will show an overall negative charge if we simply write the formula as $Ca_x^{2+}Zr_{1-x}^{4+}O_2$. One simple way for the crystal to compensate for the extra negative charge is to arrange for some of the anion sites to be vacant. For exact neutrality, the number of vacancies on the anion sub-lattice needs to be exactly the same as the number of calcium ions in the structure. Thus, each Ca^{2+} added to the ZrO_2 produces an oxygen vacancy at the same time. The formula of the crystal is $Ca_x^{2+}Zr_{1-x}^{4+}O_{2-x}$. This model is a good starting point for the discussion of the defect structure of calcia-stabilized zirconia and we will refer to it on a number of occasions later in this book.

It has been found that when a material is 'doped' with substitutional impurity cations of lower charge, anion vacancies are a common method of achieving charge balance.

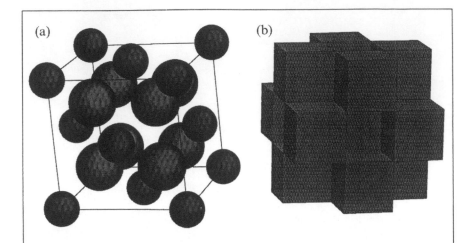

(a)   (b)

Figure 4.5 The crystal structure of CaF_2, fluorite. In (a) the fluorine atoms, larger spheres, are in a cubic array; the calcium atoms occupy half of the cube centres formed. In (b) the structure is shown as a packing of CaF_8 cubes.

The fluorite structure

The *fluorite* structure is adopted by the mineral fluorite, CaF_2, and is shown in Figure 4.5. The unit cell is cubic with a lattice parameter of 0.545 nm and contains four Ca^{2+} and eight F^- ions. In this structure, each calcium ion is at the centre of a cube of eight fluorine ions and each fluorine ion is at the centre of a tetrahedron of calcium ions. As there are twice as many fluorine atoms as metal atoms, half of the F^- cubes will be empty.

4.5.3 Interstitial formation

The two previous examples of substitution produced results rather similar to Schottky defects in the crystal. However, the anion and cation vacancies were no longer balanced, but adjusted to keep the crystal electrically neutral overall. One might expect, therefore, that substitution can also lead to unbalanced defect populations of interstitial ions which are rather similar to Frenkel defects.

This use of interstitials is quite common in materials with the *fluorite* structure shown in Figure 4.5. If CaF_2 is reacted with LaF_3, YF_3, ThF_4 or similar fluorides, non-stoichiometric phases form. In these, the metal atoms substitute for calcium on the metal ion sub-lattice in a similar way to that described in the case of calcia-stabilized zirconia. However, charge balance is ensured by the incorporation of additional F^- ions into the crystals as interstitials which occupy the vacant cube centres.

Perhaps one of the most studied oxides, because of its technological importance as a fuel, is the oxygen-rich form of uranium dioxide, UO_{2+x}. The parent phase, stoichiometric UO_2, also adopts the *fluorite* structure, with each uranium atom located at the centre of a cube of eight oxygen atoms. The composition range in UO_{2+x} is due to the additional oxygen atoms arranged at random in the normally unoccupied interstitial positions. The charge excess which will arise due to these extra oxygen ions is balanced by an equal number of U^{4+} ions becoming U^{6+} ions. In effect, this can be regarded as substitution of U^{4+} by U^{6+} together with a compensating population of oxygen interstitials. Although this explanation of the non-stoichiometry in UO_2 is largely correct, improved methods of structural analysis have shown that the interstitial oxygen atoms are not placed at random in the structure at all. In fact they are arranged in local clusters with quite specific geometries, which will be described in more detail in chapter 9.

4.6 Density and defect type

An X-ray powder photograph yields a measurement of the average unit cell dimensions of a material, and, for a non-stoichiometric compound, this invariably changes in a regular way across the phase range. In a similar way, the density of a material gives the average amount of matter in a large volume of material, and for a non-stoichiometric phase this also varies across the phase range. These two techniques can be used in conjunction with each other to determine the most likely point defect model to apply to a material. As both techniques are averaging techniques they say nothing about the real organization of the defects, but they do suggest first approximations. We will illustrate the method by reference to two typical examples, iron monoxide, with a composition close to FeO and calcia-stabilized zirconia, referred to earlier in this chapter.

4.6.1 Iron monoxide, wüstite, ~FeO

Iron monoxide, often known by its mineral name of *wüstite*, is found to possess a range of compositions close to $FeO_{1.0}$. It has a *rock salt* structure with a lattice parameter, a_o, of about 0.43 nm which varies as the composition changes. Data for a number of samples across the composition range are listed in Table 4.2. These results show that there is more oxygen present than iron in the compound. We can consider, as an initial step, two possible point defect models to account for this finding.

Table 4.2 Experimental data for ~FeO†

O:Fe ratio	Fe:O ratio	Lattice parameter (nm)
1.058	0.945	0.4301
1.075	0.930	0.4292
1.087	0.920	0.4285
1.099	0.910	0.4282

†The data here and in Table 4.3 are classical data from the paper of E.R. Jette and F. Foote, *J. Chem. Phys.* **1**, 29 (1933).

Model A. In this model we assume that the iron atoms in the crystal are in a perfect array, identical to the metal atoms in *rock salt*. In this case, to obtain an excess of oxygen, we need *interstitial oxygen atoms* to be present, as all the normal anion positions are already occupied. This possibility is shown in Figure 4.6(a). The unit cell of the *rock salt* structure contains four M atoms and four X atoms. Hence, in this model, the unit cell must contain four atoms of Fe and $4(1 + x)$ atoms of oxygen; that is, the unit cell contents are $Fe_4O_{4 + 4x}$ and the composition is $FeO_{1 + x}$, where $(1 + x)$ is the figure given in the first column of Table 4.2.

Model B. On the other hand, we could assume that the oxygen array is perfect and each oxygen atom occupies a site equivalent to that of each non-metal atom in the *rock salt* structure. As we have more oxygen atoms than iron atoms, we must, therefore, have some *vacancies in the iron positions*, as shown in Figure 4.6(b). In this case, one unit cell will contain four atoms of oxygen and $(4 - 4x)$ atoms of iron. The true formula of ~FeO now should be written $Fe_{1-x}O$.

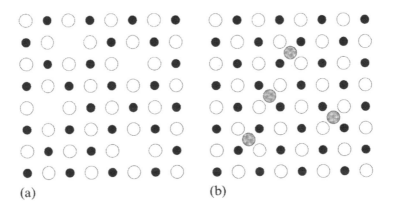

(a)   (b)

Figure 4.6 Idealized diagrams of possible defect structures of ~FeO. In (a) the shaded circles are oxygen interstitials. The Fe ions are shown as filled circles and the oxygen ions as open circles.

It is easy to determine which of these suppositions is correct by comparing the real and theoretical density of the material. For example, consider the sample specified in the top line of Table 4.2. The volume, v, of the cubic unit cell is given by $a_o{}^3$. In this case the volume is thus $(0.4301 \times 10^{-9})^{-3} \, \text{m}^3$, so $v = 7.9562 \times 10^{-23} \, \text{cm}^3$.

The mass of a unit cell is readily calculated with the knowledge that the relative atomic masses of Fe and O are 55.85 and is 16.00, respectively, and that these values correspond to the weight in grammes of Avagadro's number, N_A, of atoms. Hence

Model A. The mass of 1 unit cell is m_A,

$$m_A = \frac{[(4 \times 55.85) + (4 \times 16 \times (1 + x))]}{N_A}$$

The value of $(1 + x)$, from Table 4.2 column one is 1.058, so that

$$m_A = \frac{[(4 \times 55.85) + (4 \times 16 \times (1 + 1.058))]}{N_A} \quad \text{grammes}$$

The density, ρ, is given by the mass of one unit cell divided by the volume, to yield

$$\rho = \frac{[(4 \times 55.85) + (4 \times 16 \times 1.058)]}{N_A \times (7.9562 \times 10^{-23})}$$

$$= 6.074 \, \text{g cm}^{-3}$$

Model B. The mass of one unit cell is m_B where

$$m_B = \frac{[(4 \times (1 - x) \times 55.85) + (4 \times 16)]}{N_A}$$

and taking the value of $(1 - x)$ as 0.945, from the second column of Table 4.2, we find

$$m_B = \frac{[(4 \times 0.945 \times 55.85) + (4 \times 16)]}{N_A} \quad \text{grammes}$$

The density, ρ, is given by the mass of 1 unit cell divided by the volume, to yield

$$\rho = \frac{[(4 \times 0.945 \times 55.85) + (4 \times 16)]}{N_A \times (7.9562 \times 10^{-23})}$$

$$= 5.740 \, \text{g cm}^{-3}$$

The difference in the two values is surprisingly large, and is well within the accuracy of density determinations. The value found experimentally is $5.728\,g\,cm^{-3}$, in good accord with Model B, which assumes vacancies on the iron positions. This indicates that the formula of the $\sim$FeO phase should be written $Fe_{0.945}O$. Table 4.3 is an expanded version of Table 4.2 which includes the density data. All results are seen to be in good agreement with an experimental formula for $\sim$FeO of $Fe_{1-x}O$, in which there are vacancies at some of the Fe positions.

4.6.2   Calcia-stabilized zirconia

In the discussion of calcia-stabilized zirconia given above, it was suggested that the Ca^{2+} ions substituted for Zr^{4+} ions on normal cation sites and to maintain charge balance vacancies were introduced into the anion sub-lattice. The formula for the compound was $Zr_{1-x}Ca_xO_{2-x}$. However, we have another possibility to consider, that of Ca^{2+} cation interstitials. The simplest model to envisage is one where the Ca^{2+} interstitials are balanced by extra interstitial oxide ions, to give a formula of $ZrCa_xO_{2+2x}$. However, it is difficult to fit both Ca^{2+} and O^{2-} interstitials into the unit cell and a more plausible suggestion is to introduce one Zr^{4+} cation vacancy for each two Ca^{2+} interstitial ions present, to give a formula $Zr_{1-x/2}Ca_xO_2$.

To check these models we measure the density of the material and compare it with the theoretical density of the phase. The structure of the parent material is cubic, with a unit cell edge of about 0.52 nm. Each unit cell contains four normally occupied cation sites and eight normally occupied anion sites, so that the overall composition of stoichiometric zirconia is ZrO_2.

Turning to some experimental results, it has been found that a crystal prepared by heating 85 mole % ZrO_2 with 15 mole % CaO at 1600 °C yielded a material with a cubic unit cell which had a lattice parameter, a_o, of 0.5144 nm and a measured density of $5.525\,g\,cm^{-3}$. Let us see how these values compare with our models.

Table 4.3 Experimental and theoretical densities of $\sim$FeO

| | | | Density ($g\,cm^{-3}$) | | |
O:Fe ratio	Fe:O ratio	Lattice parameter (nm)	Observed	Interstitial oxygen	Iron vacancies
1.058	0.945	0.4301	5.728	6.076	5.740
1.075	0.930	0.4292	5.658	6.136	5.706
1.087	0.920	0.4285	5.624	6.181	5.687
1.099	0.910	0.4282	5.613	6.210	5.652

Model A. $Zr_{1-x}Ca_xO_{2-x}$ with a*nion vacancies.* The mass of the unit cell of this material will be given by

$$m = \frac{[4 \times (0.85 \times 91.22) + 4 \times (0.15 \times 40.08) + 4 \times (1.85 \times 16)]}{N_A}$$

$$= \frac{[452.6]}{N_A}$$

where 91.22 is the molar mass of Zr, 40.08 is the molar mass of Ca, 16 is the molar mass of O and N_A is Avagadro's number. The volume, *v*, of the unit cell will be a_o^3, that is

$$v = (5.144 \times 10^{-8})^3 \text{ cm}^3$$

so that the density, ρ, is found to be

$$\rho = 5.522 \text{ g cm}^{-3}$$

Model B: $Zr_{1-(x/2)}Ca_xO_2$ with Ca^{2+} *interstitials.* The mass of the unit cell is

$$m = \frac{[4 \times (0.925 \times 91.22) + 4 \times (0.15 \times 40.08) + 4 \times (2.0 \times 16)]}{N_A}$$

Where all the symbols have the same meaning as before. Additionally, the volume of the unit cell will be unchanged, so that the density is found to be

$$\rho = 5.972 \text{ g cm}^{-3}.$$

The calculations come down squarely on the side of the substitution plus anion vacancy model.

The experimental data are shown in a more extended form in Figure 4.7. The calculated values for the density of the samples has been shown for both the vacancy and interstitial models that we have just described. As in the case of wüstite, these densities are sufficiently different to discriminate between the two models over the range of compositions studied. Also shown are the experimentally determined densities for two series of samples, one prepared at 1600 °C and the other at 1800 °C. The results are interesting. At 1600 °C, the vacancy model fits the data pretty well. In the samples prepared at the higher temperature, however, the situation is more complex. For low amounts of CaO, it seems that we have interstitials present. At greater concentrations of CaO, however, the situation is less clear and it would appear that we pass through a region in which some cells seem to contain interstitials, while in others the substitution mechanism is still employed.

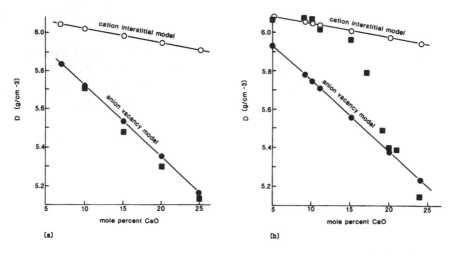

Figure 4.7 A comparison of observed and calculated densities for calcia-stabilized zirconia. Here (a) shows samples quenched from 1600 °C and (b) shows samples quenched from 1800 °C. The calculated density values are shown as open circles for a cation interstitial model, and as filled circles for an anion vacancy model. The experimental densities are shown as filled squares. [Data redrawn from A.M. Dienes and R. Roy, *Solid State Comm.* 3, 123 (1965).]

4.7   Interpolation

Interpolated atoms are atoms which occupy normally empty positions in a crystal structure, and so can be considered to be similar to Frenkel defects. The likelihood of finding that a non-stoichiometric composition range is due to the presence of interpolated atoms in a crystal will depend on the openness of the structure and the size of the impurity. Despite this limitation, non-stoichiometric materials which utilize an interstitial mechanism are many and varied. Examples are: *interstitial alloys*, in which small atoms percolate into a metallic host; layered structures where atoms are taken in between weakly held layers, typified by TiS$_2$; and structures containing *tunnels*, for instance the cubic *tungsten bronzes*.

4.7.1   *Interstitial alloys and hydrides*

The *interstitial alloys* are formed when small atoms such as C or N fit into the spaces between larger metal atoms such as Fe. These alloys are extremely hard and metallic components are often given a wear-resistant coating of an interstitial alloy by diffusing C or N into the surface layers. Metal *hydrides* are another group of materials formed by interpolation. In these deceptively simple non-stoichiometric compounds, hydrogen diffuses into the structure of a host metal to form a variety of ordered and disordered phases.

4.7.2 Titanium disulphide

A number of very interesting compounds form when a structure with layers incorporates small or medium-sized electropositive ions such as Li^+. The resulting compounds, often called *insertion compounds*, are finding increasing use in batteries, sensors and displays, and examples of these uses will be given in later chapters.

One important group of non-stoichiometric materials are derived from disulphides with a layered form of structure, such as TiS_2 and NbS_2. To illustrate this behaviour, let us look at the titanium sulphides. The structure

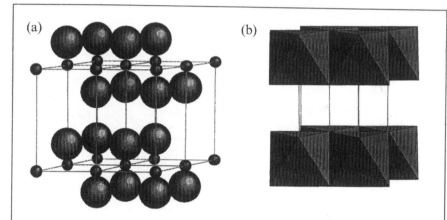

Figure 4.8 The structure of CdI_2 (a) as a perspective view and (b) as a packing of CdI_6 octahedra. In both structures, the iodine atoms are arranged in a hexagonal stacking. In CdI_2, the octahedral sites between alternate sheets of iodine atoms are filled by cadmium atoms.

The CdI_2 structure

The CdI_2 structure type is shown in Figure 4.8. The unit cell is hexagonal with parameters $a = 0.424\,\text{nm}$ and $c = 0.684\,\text{nm}$. It is made up of layers of iodine atoms stacked up in a hexagonal fashion as shown in Figure 4.8(a). This sort of stacking leaves octahedral sites between the layers. In CdI_2, every other sheet of these octahedral sites is filled by cadmium to generate layers of composition CdI_2 as emphasized in Figure 4.8(b). These layers are only weakly held together by secondary bonding. CdI_2 is easily cleaved into sheets parallel to the layers as a result.

of TiS_2 is of the CdI_2 type, shown in Figure 4.8. It is made up of layers of composition TiS_2 which are only weakly held together.

Small atoms such as Li can enter the structure in varying amounts to form non-stoichiometric phases with a general formula Li_xTiS_2. Because the bonding between the layers is weak, this process is easily reversible and the compound acts as a convenient reservoir of Li atoms. The usefulness of this material will be discussed in chapter 5.

Another interesting series of non-stoichiometric phases can be generated by gradually filling the vacant octahedral sites in TiS_2 with Ti itself. Depending on the preparation conditions used, the interpolated atoms may

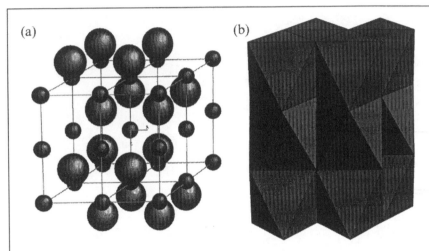

Figure 4.9   The structures of NiAs (a) as a packing of atoms and (b) as a packing of $NiAs_6$ octahedra. In both structures the As atoms are arranged in a hexagonal stacking. Ni atoms occupy all of the octahedral sites formed between these layers.

The NiAs structure

The NiAs structure is the hexagonal analogue of the cubic NaCl structure. The hexagonal unit cell has parameters $a = 0.360\,nm$, $c = 0.501\,nm$. The structure is shown as a packing of atoms in Figure 4.9(a). Each Ni atom is surrounded by an octahedron of As atoms and each As atom by an octahedron of Ni atoms. This arrangement is emphasized in Figure 4.9(b). It is simply related to the CdI_2 structure by the filling of the interlayer octahedral sites by metal atoms.

be ordered, disordered or partially ordered. If they are ordered, X-ray diffraction will record the presence of new phases. In this way a number of intermediate compounds have been recognized, among which are Ti_8S_9, Ti_4S_5, Ti_3S_4, Ti_2S_3 and Ti_5S_8. Ultimately, all of the available octahedral sites are filled and the material has a composition TiS, with the nickel arsenide, NiAs, structure.

4.7.3 Cubic tungsten bronzes

Good examples of interpolation are provided by the fascinating compounds known as tungsten 'bronzes'. The tungsten bronzes are so called because

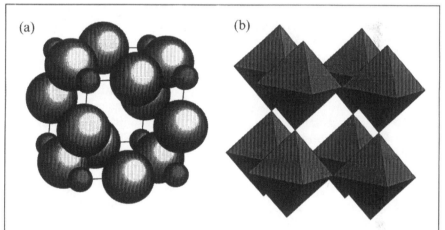

Figure 4.10 The structure of WO_3 shown (a) as a perspective view and (b) as an array of corner-sharing WO_6 octahedra. The large cage sites between the octahedra take in interpolated metal atoms in the cubic tungsten bronzes.

The WO_3 structure

The structure of WO_3 is made up of W^{6+} ions surrounded by octahedra of O^{2-} ions. The unit cell is orthorhombic at room temperature, due to small distortions of the WO_6 octahedra. The ideal cubic unit cell has a lattice parameter 0.38 nm. The structure as a packing of atoms is shown in Figure 4.10(a). The structure is often more conveniently viewed as a three-dimensional chessboard-like arrangement of corner-linked WO_6 octahedra, as shown in Figure 4.10(b).

when they were first discovered, by Wohler in 1837, their metallic lustre and conductivity led him to believe that he had made some new alloys of tungsten rather than new oxides. The lithium tungsten bronze Li_xWO_3 has a composition range from WO_3 to $Li_{0.5}WO_3$ and is formed by interpolation of the rather small Li atoms at random into the rather open parent WO_3 structure. The open nature of the tungsten trioxide structure allows the lithium atoms to move readily in and out of the crystals, and like Li_xTiS_2, Li_xWO_3 can act as a convenient solid reservoir for Li metal.

4.8   Defect chemistry

Point defect populations profoundly affect both the physical and chemical properties of materials. In order to describe these consequences, we need a notation for defects that is simple and self-consistent. The most widely employed system, and the one we shall use in this book, is the Kröger–Vink notation.

One of the most difficult problems, when working with defects in ionic crystals, is to decide on the charge on the ions and atoms of importance. In the Kröger–Vink notation, this problem is overcome in the following way. When we add or subtract elements to or from a crystal we do so by adding or subtracting electrically neutral atoms. When ionic crystals are involved, this requires that we separately add or subtract electrons. To illustrate the implications of this idea we will use the notation to describe some defects in a compound of formula MX, where M is a metal and X an anion. We commence with a consideration of uncharged atomic defects.

4.8.1   Atomic defects

Vacancies. When empty lattice sites occur, they are indicated by the symbols V_M and V_X for the metal and non-metal sites, respectively. For example, in an ionic oxide such as NiO, V_{Ni} would imply the removal of a Ni^{2+} ion together with two electrons, that is, a neutral Ni atom. Similarly, V_O would indicate a vacancy in the oxygen sub-lattice and implies removal of an O^{2-} ion from the crystal and the subsequent addition of two electrons to the crystal.

Interstitial atoms. When atoms occupy interstitial positions, they are denoted by M_i and X_i for metals and anions, respectively. Hence K_i represents an interstitial potassium atom in a crystal.

Impurity atoms. Many materials contain impurity atoms, either introduced on purpose, or because purification procedures are inadequate, and it is important to be able to specify the nature of the impurities and where in the crystal they are to be found. This is particularly true for dopant impurities that are deliberately added to control electronic or other properties. In this case, the impurity is given its normal chemical symbol and the site occupied is written as a subscript. Thus, an Mg atom on a Ni site in NiO would be written as Mg_{Ni}. The same nomenclature is used if an atom in a crystal occupies the wrong site. So it is possible for M atoms to be on X sites, written as M_X or X atoms to be on M sites, written as X_M. A potassium atom on a bromine site in KBr would be written as K_{Br}, for example.

Associated defects. It is possible for one or more lattice defects to associate with one another, that is, to cluster together. These are indicated by enclosing the components of such a cluster in parenthesis. As an example $(V_M V_X)$ would represent a Schottky defect in which the two vacancies were associated as a vacancy pair.

It can be noted that the nomenclature uses a straightforward system of description. It is seen that the normal symbol for a chemical element represents the species involved, and the subscript represents the position of the atom in the structure. Apart from the symbols for chemical elements, we have used V to represent a vacancy and i to represent an interstitial. The symbol V is also the chemical symbol for the element vanadium, of course. Where confusion may occur, the symbol for a vacancy is written Va.

4.8.2 Charges on defects

Electrons and electron holes. The charged defects that most readily come to mind are electrons. Some fraction of the electrons in a crystal may be free to move through the crystal. These are denoted by the symbol e′. The superscript ′ represents the negative charge on the electron and it is written in this way to emphasize that it is considered relative to the surroundings rather than as an isolated point charge. This becomes important when we are concerned with the interactions and reactions of defects. Although electrons are the only charged sub-atomic particles to exist in the structure, it often simplifies matters to think about the sites where electrons are missing. This is analogous to thinking about vacancies instead of atoms. In the case of these 'electron vacancies' we use the symbol h˙ to denote the defect, which is called an *electron hole*, or, more commonly, simply as a '*hole*'. Each hole will bear a positive charge of $+1$, which is represented by the superscript dot to emphasize that it is considered relative to the surrounding lattice.

Charges on defects. Besides the electrons and holes just mentioned, the atomic defects that we have described above can also carry a charge. In ionic crystals, this may be considered to be the normal state of affairs. The Kröger–Vink notation bypasses the problem of deciding on the real charges on defects by considering only effective charges on defects. The *effective charge* is the charge that the defect has with respect to the normal crystal lattice. To illustrate this concept, let us consider the situation in an ionic material such as NaCl, which we will supposed to be made up of the charged ions Na^+ and Cl^-. If we have a vacancy in the NaCl structure at a sodium position V_{Na}, what will the effective charge on this defect be? To answer this, you must mentally think of yourself as 'diffusing' through the NaCl structure. Each time an Na^+ ion is encountered, a region of positive charge will be experienced. If, then, we meet a vacancy instead of a normal ion, this will not seem to be positive at all. Relative to the situation normally met with at the site, we will encounter a region which has an effective negative charge, that is a charge relative to that normally encountered at that position equivalent to -1. In order to distinguish effective charges from real charges, the superscript prime is used for each unit of negative charge and the superscript dot is used for each unit of positive charge. Hence a 'normal' vacancy at a sodium site in NaCl would be written as V'_{Na}, which corresponds to a missing Na^+ ion. Similarly, a 'normal' vacancy at a chloride ion site would seem to be positively charged relative to the normal situation in the crystal. Hence the vacancy has an effective charge of $+1$, which would be written $V^{\bullet}_{Cl}$.

With each of the other defect symbols V_M, V_X, M_i, M_X and associated defects such as $(V_M V_X)$ an effective charge relative to the host lattice is also possible. Thus, $Zn_i^{2\bullet}$ would indicate a Zn^{2+} ion at an interstitial site which is normally unoccupied and hence without any pre-existing charge. In such a case, all the charge on the Zn^{2+} ion is experienced as we move through the lattice, and hence the presence of two units of effective charge is recorded in the symbol, i.e. $2^{\bullet}$. Similarly, substitution of a divalent ion such as Ca^{2+} for monovalent Na^+ on a sodium site gives a local electronic charge augmented by one extra positive charge which is then represented as $Ca^{\bullet}_{Na}$.

Suppose now a sodium ion in NaCl, represented by Na_{Na}, is substituted by a potassium ion, represented by K_{Na}. Clearly the defect will have no effective charge, as, to anyone moving through the crystal, the charge felt on encountering the K_{Na} ion is the same as that experienced on encountering a normal Na_{Na} ion. This defect is, therefore, neutral in terms of effective charge. This is written as K^x_{Na} when the effective charge situation needs to be specified. The superscript x represents no effective charge at the site in question.

It is, therefore, evident that the idea of the charge on the defect is separated from the chemical entity which makes up the defect. Real charges are represented by $n+$ and $n-$, while effective charges are represented by $n\prime$

and $n\bullet$, or x. It is for this reason that the charges on electrons and electron holes mentioned above were written as $'$ and $\bullet$, as these charges are only of importance relative to the surrounding crystal lattice.

The main features of the Kröger–Vink notation are summarized in Table 4.4.

4.8.3 Reaction equations

There are many instances when we have to consider reactions which cannot be expressed within the normal chemical nomenclature. For example, if an impurity is doped into a crystal it can have profound effects on the physical and chemical properties of the substance because of the defects introduced. However, defects do not occur in the balance of reactants expressed in traditional chemical equations and so these important effects are lost to the chemical accounting system that the equations represent. If defects can be incorporated into normal chemical equations, it will not only allow us to keep a strict account of these important entities but also to apply chemical thermodynamics and other techniques of handling chemical energy exchange to the reactions. We can, therefore, build up a *defect chemistry*, in which the defects play a role analogous to that of the chemical atoms themselves. The Kröger–Vink notation allows this to be done, provided the normal rules which apply to balanced chemical equations are preserved. As the rules are slightly different than those of elementary chemistry they are set out here.

Table 4.4 Summary of the Kröger–Vink notation†

Defect Type	Notation	Defect type	Notation
Non-metal vacancy at non-metal site	V_X	Impurity non-metal (Y) at non-metal site	Y_X
Metal vacancy at metal site	V_M	Impurity metal (A) at metal site	A_M
Neutral vacancies	$V_M^x\ V_X^x$	Non-metal vacancies with positive effective charge	$V_X^\bullet$
Metal vacancies with negative effective charge	V_M'	Interstitial metal	M_i
Interstitial non-metal	X_i	Interstitial metal with positive effective charge	$M_i^\bullet$
Interstitial non-metal with negative effective charge	X_i'		
Free electron‡	e'	Free positive hole‡	$h^\bullet$

†The definitive definitions of this nomenclature and further examples are to be found in the IUPAC *Red Book on the Nomenclature of Inorganic Chemistry*, Chapters 1–6.
‡Concentrations of these defects are designated by n and p, respectively.

1. *The number of metal atom sites must always be in the correct proportion to the number of non-metal atom sites in the crystal.* For example, in MgO we must always have equal numbers of both types of position; in TiO_2, there must always be twice as many anion sites as cation sites; and in a compound M_aX_b, there must be *a* metal atom sites for every *b* non-metal atom sites. As long as this proportion is maintained, the *total* number of sites can vary, as this simply corresponds to more or less substance present. If the crystal contains vacancies these must also be counted in the total number of sites, as each vacancy can be considered to occupy a site just as legally as an atom. Interstitial atoms do not occupy normal sites and so do not count when this rule is being applied.

2. *The total number of atoms on one side of the equation must balance the total number of atoms on the other side.* Remember that the subscripts and superscripts are labels describing charges and sites, and are not counted in evaluating the atom balance.

3. *The crystal must always be electrically neutral.* This means not only that the total charge on one side of the equation must be equal to the total charge on the other side, but also that the sum of the charges on each side of the equation must equal zero. In this assessment, both effective and real charges must be counted if both sorts are present.

4. *When crystals react, only neutral atoms are involved.* After the reaction, neutral atoms can dissociate into charged species if this is thought to represent the real situation in the crystal.

To illustrate exactly how these rules work and to show that their application is not difficult, let us consider the reactions that can occur when crystals of ZrO_2 are reacted with CaO to produce a crystal of calcia-stabilized zirconia. This sort of situation cannot be treated by normal chemical equations, but it is clear that such reactions do take place and are important.

In ZrO_2 there are twice as many anion sites as there are cation sites. Let us suppose, as we have in the past, that the reacting Ca atoms are located on normal cation sites. In order to comply with the first rule we must, therefore, create two anion sites per Ca atom. These are considered to be vacant at the start. However, as we have to locate an oxygen atom in the crystal as well, it is reasonable to place it in one of these sites. The other site remains vacant. We have noted that the reactions are to be carried out using neutral atoms so as to avoid mistakes over the allocation of charges to reacting species. The reaction equation is then

$$CaO \xrightarrow{ZrO_2} Ca_{Zr}^{4\prime} + V_O^{2\bullet} + O_O^{2\bullet}$$

This means that in the structure of ZrO_2 we now have Ca atoms and O atoms on sites normally occupied by metal and non-metal species. As the Ca atom is taken as being neutral, the effective charge at the site will be $4\prime$ with respect to the charge encountered when a normal Zr ion is encountered.

Similarly, the oxygen atom will be neutral and so the effective charge on the oxygen site which is occupied will be 2•, and that on the oxygen site which is vacant will equally be 2•. Note that the equation conserves mass balance, electrical charge balance and site numbers, in accordance with the rules given above.

Now, ZrO_2 and CaO are normally regarded as ionic compounds, so that ions should occupy the sites, not neutral atoms. This gives us an alternative and perhaps more realistic process which we can write as

$$CaO \xrightarrow{ZrO_2} Ca_{Zr}^{2\prime} + V_O^{2\bullet} + O_O^x$$

This equation also conserves mass balance, electrical charge balance and site numbers, as indeed it must. We notice that the effective charge on the Ca ion is now $2\prime$ as the normal charge at a cation position is due to the presence of Zr^{4+} ion, and so, with respect to the normal situation, the presence of the Ca^{2+} ion leads to an effective decrease in the charge encountered at the site in question by two units. Similarly, the oxygen ion occupies a normal oxygen ion site and we have no difference from that normally encountered in ZrO_2 and so the effective charge for these ions is zero.

It may be argued that the Ca^{2+} ions do not occupy Zr sites, but prefer interstitial positions. The Ca^{2+} ions are then easy to deal with, as they do not affect the site numbers in the ZrO_2 matrix. Each oxygen atom can again be assumed to occupy an anion site. Now in this case, the site conservation rule applies, and for each fresh anion site created we must create one half of a new cation site. This simply means that every two oxygen atoms incorporated into the crystal generate one new cation position. As the Ca ions do not make use of these positions they remain empty. Once again, taking the atomic entities to be ions rather than neutral atoms, we can write the formation equation as

$$2CaO \xrightarrow{ZrO_2} 2Ca_i^{2\bullet} + 2O_O^x + V_{Zr}^{4\prime}$$

So far we have written down three equations which could apply to the reaction of CaO with ZrO_2 to form calcia-stabilized zirconia. All of them are correct in a chemical sense. To decide which of them, if any, represents the true situation in the material, experimental evidence, such as that derived from density measurements of the type already shown in Figure 4.11, is needed.

4.9 Point defect interactions

We have seen that defects in a crystal can carry effective charges, and because of this we would expect the defects to interact with each other quite strongly. It is worthwhile to see if we can gain some approximate feeling for the magnitude of these energy terms. As we are thinking about the interaction of charged defects, perhaps the place to start is with simple electrostatic theory. This gives the energy of interaction of two unit charges (sometimes expressed as the work needed to separate them) as

$$E_{electro} = \frac{e^2}{4\pi\varepsilon_0 r}$$

where each of the charges has a magnitude of e and we assume that the charges have opposite signs and attract each other, r is the separation of the charges and ε_0 is the permittivity of vacuum.

If we apply this formula to defects in a crystal, and again assume that the defects are oppositely charged, so that they attract each other, the energy term will be roughly equivalent to the enthalpy of formation of a defect pair, ΔH_p. In order to allow for the crystal structure itself, which will modify the interaction energy considerably, we make the assumption that the force of attraction is simply 'diluted' in the crystal by an amount equal to its relative permittivity.

The modified formula is then

$$\Delta H_p = \frac{(z_1 e)(z_2 e)}{4\pi\varepsilon\varepsilon_0 r}$$

where ΔH_p is the enthalpy of interaction, z_1 and z_2 are the effective charges on the defects, ε is the static relative permittivity of the crystal and the other symbols have the same meaning as in the initial formula.

Simple as this theory is, it is good enough to tell us whether the association of defects is likely to occur. Consider, as an example, a Schottky defect, consisting of a cation vacancy and an anion vacancy, in a crystal of a monovalent metal MX with the *rock salt* structure. These vacancies will have effective charges of $+e$ and $-e$. Their interaction will be greatest when they are closest to each other, that is, when they occupy neighbouring sites in the crystal. The separation of these sites is about 3×10^{-10} m in this structure. An approximate value for the relative permittivity of a *rock salt* structure crystal is 10. The value of ε_0 is given by 8.854×10^{-12} F m^{-1} and the electronic charge by 1.6022×10^{-19} C, so the interaction energy, which is attractive, is given by

$$\Delta H_p = \frac{(1.6022 \times 10^{-19})^2}{4 \times 3.1416 \times 10 \times 8.854 \times 10^{-12} \times 3 \times 10^{-10}}$$
$$= 7.69 \times 10^{-20} \text{ J.}\dagger$$

†The units in this calculation work out correctly.

$$\Delta H_p = \frac{C^2}{F\,m^{-1}m} = \frac{C^2}{F}$$

Converting into more fundamental values C = As, F = C V^{-1}, V = W A^{-1} and W = J s^{-1}, where C = coulomb, A = ampere, F = farad, V = volt, W = watt and J = joule. Making the appropriate substitutions,

$$\frac{C^2}{F} = CV = As\,J\,s^{-1}A^{-1} = J$$

The figure we have calculated is the interaction energy for one pair of vacancies only. To obtain the molar quantity, we multiply ΔH_p above by Avagadro's number, N_A, to yield

$$\Delta H_p = 46.3 \, \text{kJ mol}^{-1}$$

This is similar in magnitude to the values quoted for Schottky defect formation energies and so we would expect that a reasonable proportion of the vacancies would be associated into pairs.

We can actually make an assessment of the fraction of defects in a crystal which are associated using the rough interaction energies calculated above in the following way. The Boltzmann equation tells us that if we have two energy states separated by an energy difference ΔH the fraction of the population in the upper state, f, is given by

$$f = \exp\left(-\frac{\Delta H}{kT}\right)$$

where k is Boltzmann's constant and T the absolute temperature. We can use this for our purposes, and to provide an example, let us return to the case discussed above, that of vacancy pair association in an $NaCl$ type material.

If the number of Schottky defects is n_s, we will have n_s cation vacancies and n_s anion vacancies in the crystal. If we take the interaction energy to be 7.69×10^{-20} J, as we calculated, the fraction of vacancies associated will be given by f, where

$$f = \exp\left[\frac{\left(-7.69 \times 10^{-20}\right)}{\left(1.38 \times 10^{-23} \times 10^3\right)}\right]$$

where we have taken a value of $1.38 \times 10^{-23} \, \text{J K}^{-1}$ for Boltzmann's constant, k, and a temperature of 1000 K. Hence

$$f = 0.0038$$

That is, about four defects in every 1000 will be associated into pairs at 1000 K. As we already know how to estimate the number of Schottky defects in a crystal, it is, therefore, possible to find the total number of vacancies that are associated in pairs. Obviously we can use similar reasoning to that above for other defect types.

We can conclude this section by observing that although the estimates of interaction energy given here are approximate, they do suggest that a fair number of defects will exist as clusters rather than as isolated 'point defects'. This conclusion has been borne out in recent years by realistic calculations which can be made via a variety of sophisticated computational techniques. These show that much of the interaction energy between point defects arises when the atoms in the crystal close to the defects move slightly to adjust to the new configuration of the defects themselves. This process is known as *relaxation*, and this relaxation energy makes a major contribution to the overall stabilization energy of defect clusters.

4.10 Supplementary reading

There are very few books which cover the defect chemistry and physics covered in this chapter at an introductory level. Probably the most complete account of point defect chemistry, and an explanation of the Kröger–Vink notation is to be found in the book:

F.A. Kröger, *The Chemistry of Imperfect Crystals*, 2nd edition, North–Holland, Amsterdam (1974).

An account of the defect chemistry of oxides with much experimental data can be found in:

P. Kofstad, *Non-stoichiometry, Diffusion and Electrical Conductivity in Binary Metal Oxides*, Wiley–Interscience, New York (1972).

A brief but interesting account of defect chemistry is given in chapters 1 and 5 of:

W.J. Moore, *Seven Solid States*, Benjamin, New York (1967).

Two very good review articles on the subject matter of this chapter are:

J.S. Anderson, in *Chemistry of the Solid State*, ed. C.N.R. Rao, Marcel Dekker, New York (1974).
D.J.M. Bevan, *Comprehensive Inorganic Chemistry*, Vol. 4, chapter 49, ed. A.F. Trotman-Dickenson, Pergamon, Oxford (1973).

5 Fast ion conductors

5.1 Introduction

Batteries are obviously very useful. The key to battery construction is an *electrolyte* which can carry ionic conduction but not electronic conduction. The first batteries used water solutions as electrolytes and such batteries are still common place. The normal lead–acid car battery is a good example. However, batteries which utilize a solid electrolyte have been developed and find uses in a wide range of applications from watches to heart pacemakers, where a liquid electrolyte battery would not be satisfactory. Moreover, the voltage produced by a battery depends on the concentration of the materials on either side of the electrolyte. This means that we can use the voltage as a signal to give us concentration information. This forms the basis of the operation of many *sensors*. If the battery voltage is reversed, as it is when charging a car battery, then material is pumped from one side of the electrolyte to the other. Thus, batteries can be modified and used as *electrochemical pumps* capable of delivering very precisely monitored amounts of substance to one side of the cell.

With few exceptions, LiI discussed in section 5.2, for example, normal solids have too low a conductivity to be useful in these devices. In fact, the principle requirement of a solid electrolyte is that it should conduct ions through the crystal lattice as fast as ions can travel in a solution. This is quite a demanding requirement! However, careful manipulation of the defect populations present in 'normal' crystals have led to the production of a number of materials which do have ionic conductivities in the solid state which are as large as that normally found in solutions. Such materials are sometimes called *super ionic conductors*, but the term *fast ion conductors* is preferred to avoid confusion with metallic superconductors, which transport electrons, not ions, and by a quite different mechanism. Figure 5.1 shows the domains of fast ion conduction for the principle materials discussed in this chapter. This chapter focuses on the role that defects play in fast ion conductors and uses this information to explain the mechanisms by which devices which rely on these materials operate.

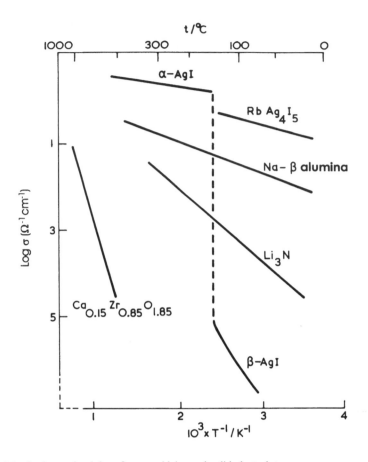

Figure 5.1   Ionic conductivity of some widely used solid electrolytes.

5.2   The lithium iodide battery

Lithium iodide is an electrolyte in certain batteries used in specialist situations. Although LiI is not a fast ion conductor, this disadvantage is more than offset by reliability and long life, which make these batteries ideal for medical use in heart pacemakers, for example. These features also make LiI cells useful in backup circuits designed to function in emergencies. For heart pacemakers, the battery itself is typically constructed of two cells, placed back to back, separated by a nickel gauze and contained in a stainless steel or titanium case, while for use in electronic circuits, a single button cell of the type shown in Figure 5.2 is more often employed. The anode is made of lithium metal. For the cathode, a conducting polymer of iodine and polyvinyl pyridine is employed because iodine itself is not a good electronic conductor. The cell is fabricated by placing the Li anode in contact with the

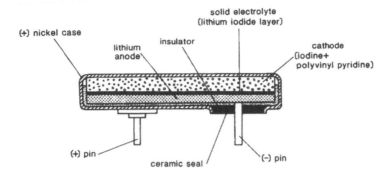

Figure 5.2 A single LiI button cell for use in electronic circuits.

polyvinyl pyridine–iodine polymer. The lithium, being a reactive metal, immediately combines with the iodine in the polymer to form a thin layer the electrolyte, LiI. The reaction is

overall reaction : $Li(s) + \frac{1}{2}I_2(s) \longrightarrow LiI(s)$

The voltage of the battery is 2.8 V.

In order for the battery to function, the LiI must be a reasonable ionic conductor. The cell operation is sustained by the Schottky defect population in the LiI. On closing the external circuit, the Li atoms in the anode surface become Li^+ ions at the anode–electrolyte interface

anode reaction : $Li(s) \longrightarrow Li^+ + e^-$

These diffuse through the LiI via the Schottky cation vacancies to reach the iodine in the cathode. The electrons lost by the Li traverse the external circuit and arrive at the interface between the cathode and the electrolyte. Here they react with the iodine and the incoming Li^+ ions to form more LiI

cathode reaction : $2Li^+ + I_2 + 2e^- \longrightarrow 2LiI$

These reactions are shown schematically in Figure 5.3. During use, the thickness of the LiI electrolyte gradually increases because of this reaction, and it is this factor that ultimately causes the cell to become unusable.

Although the Schottky defect population in LiI is vital for maintaining battery operation it is too low for many purposes. For instance, the low ionic conductivity means that the current in the external circuit cannot be large. This is due to the fact that the small number of Schottky defects present limits the magnitude of the current flow in the external circuit. To overcome this problem the LiI is sometimes doped with Ca^{2+} ions. Recall that the consequence of doping an ionic material with a cation of higher

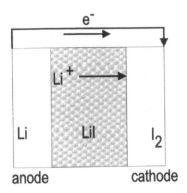

Figure 5.3 The processes taking place in a LiI cell. Electrons traverse the external circuit from anode to cathode. Li^+ ions travel across the LiI electrolyte via Schottky vacancy diffusion.

valence is the introduction of cation vacancies. So each Ca^{2+} ion in LiI will form one cation vacancy over and above those present due to Schottky defects. In this way, the conductivity of the electrolyte can be substantially increased. Nevertheless, conductivities equivalent to solutions cannot be achieved in LiI by this or any other method. In later sections we will see how materials with enhanced defect populations make it possible to achieve this goal.

5.3   Disordered cation compounds

Although most fast ion conductors have only been developed in recent years, they are not altogether new and some were investigated by Faraday in the nineteenth century. Examples of these materials are listed in Table 5.1. They show a very high ionic conductivity at temperatures above about 150 °C due to a very high concentration of what might be thought of as Frenkel defects.

The origin of these high-defect concentrations can be understood by reference to the apparently simple material, AgI. At room temperature, AgI exists in the β-form, which has the *wurtzite* structure. The ionic conductivity of this material is normal. However, in 1914 silver iodide was discovered to transform to a high temperature polymorph, α-AgI, above 147 °C. This material possesses an unusually high ionic conductivity, as can be seen from the conductivity data in Figure 5.1.

The structure of the α- phase, shown in Figure 5.4, reveals that the iodine ions form a fixed, body-centred cubic sub-lattice. The unit cell contains two AgI formula units and so two Ag^+ ions must, therefore, be distributed in some way between the iodide ions. There are quite a number of different possible cation sites available: six with octahedral geometry, 12 with

Table 5.1 Disordered cation compounds related to α-AgI†

bcc anions	fcc anions	Miscellaneous
α-AgI	α-CuI	Na_2S
α-Ag_2S	α-Cu_2Se	MHg_4I_5 (M = Rb, K, Cs)
α-Ag_2Te		α-CuAgS
α-Ag_2Se		
α-Ag_3CuS_2		
α-Ag_3SI		
Ag_3SBr		
α-Ag_3HgI_4		
$Ag_4HgSe_2I_2$		

†In these phases the α-form refers to the high temperature disordered phase, the β-form to the ordered room temperature phase and the γ-form to any additional low temperature ordered phases which may occur.

tetrahedral geometry and 24 with trigonal geometry, making 42 in all! In α-AgI the silver ions use only the tetrahedral positions so that two will be occupied and 10 empty. However, the two which are occupied are continually changing as the Ag^+ ions continually jump between all the tetrahedral sites. Over the course of several seconds, it appears that the Ag^+ ions are distributed statistically between these positions as if they were endlessly flowing from one to another. This has led to the concept of a *molten sub-lattice* of Ag^+ ions moving like a liquid through a fixed matrix of I^- ions.

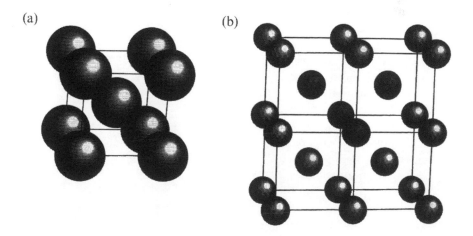

(a)　　　　　　(b)

Figure 5.4 (a) The body-centred cubic structure of the I^- ions in α-AgI. (b) Four unit cells arranged to show the tetrahedral sites, between the darker ions, occupied on a statistical basis by continually moving Ag^+ ions.

Structural studies have shown that a similar thing happens in the other materials listed in Table 5.1. At very low temperatures the structures are quite normal in the sense that all atoms can be placed in well-defined locations in the crystal structure. These low temperature structures are usually labelled the γ-form. At temperatures close to room temperature, it is found that some or all of the cations exhibit very pronounced uncertainty in their positions, due to greatly increased amplitudes of vibration. These room temperature forms are normally labelled as β-. Above a transition temperature, which is not too far from 400 K for all of these compounds, the α-phase is formed. In the high temperature forms, the cations occupy some or all positions statistically, leading to the high metal atom mobility described. Thus, the picture that emerges is one in which, as the temperature increases, metal atom vibrations increase until, at the transition temperature, these motions are so extreme that neighbouring sites become partly occupied. At this temperature the metal atoms are unable to distinguish between sites which should be permanently occupied and those which should be normally empty. If we imagine that the atoms are rapidly jumping from site to site in this high temperature phase, we have a picture of large numbers of Frenkel defects constantly forming and being annihilated.

5.4   Calcia-stabilized zirconia and related fast oxygen ion conductors

5.4.1   Structure and oxygen diffusion in fluorite structure oxides

A number of oxides with the *fluorite* structure are widely used in solid state electrochemical systems. They have formulae AO_2xCaO, where A is typically Zr, Hf and Th, or $ZrO_2xM_2O_3$ where M is usually La, Sm, Y, Yb or Sc. Calcia-stabilized zirconia is the most important material of this type. The technological importance of these materials lies in the fact that they are fast ion conductors for oxygen ions at moderate temperatures. This property is enhanced by the fact that there is negligible cation diffusion or electronic conductivity in these materials which makes them ideal for use in a diverse variety of batteries and sensors.

In order to understand how these oxides are able to conduct oxygen ions with facility it is necessary to reconsider their defect structure. The stoichiometric composition of a *fluorite* structure oxide is MO_2. Taking calcia-stabilized zirconia as an example, we know that addition of CaO drops the metal to oxygen ratio to below 2.0, and the formula of the oxide becomes $Ca_xZr_{1-x}O_{2-x}$. As we described in the previous chapter, the structure of calcia-stabilized zirconia fabricated at temperatures of about 1600 °C is one in which the Ca^{2+} ions substitute on sites normally occupied by Zr^{4+} ions and we have compensating vacancies on the oxygen sites. For each Ca^{2+} ion inserted into the structure, we must create one anion vacancy. Hence an oxide containing 20 mole % CaO will have 20 mole % oxygen vacancies in the structure. Exactly the same situation holds when zirconia is reacted with the M_2O_3 oxides mentioned above. The population of oxygen vacancies will be rather less per substituted ion in these latter compounds, as one oxygen vacancy will form for every two foreign 3+ cations incorporated, leading to a generalized formula of $M^{3+}Zr_{1-x}O_{2-x/2}$.

Nevertheless, all of these stabilized zirconias contain an enormous oxygen vacancy population. The result of this is that the diffusion coefficient of oxygen ions is increased by many orders of magnitude compared to a normal oxide. This is because every normal oxygen ion is next to an oxygen vacancy and the rather open *fluorite* structure ensures that the energy barrier to be overcome in making a jump is rather low. These materials are a very fast oxygen ion conductors with no the cation transport or electronic conductivity.

5.4.2   Stabilized zirconia electrolytes

The high defect population in these stabilized zirconias is used in a large number of surprising ways. All of them rely on the zirconia being used as a solid electrolyte but they are not all batteries in the conventional sense. The reason for this flexibility arises as a consequence of the high oxygen vacancy concentrations.

What will tend to happen if a plate of stabilized zirconia separates oxygen gas at two different pressures, as shown in Figure 5.5? The high oxygen ion diffusion coefficient will allow ions to move from the high pressure side to the low pressure side so as to even out the pressure differential. However, this process will not continue for more than a transient moment because the oxygen gas on the high pressure side needs to be ionized and the oxygen ions arriving at the low pressure side need to gain electrons to form oxygen atoms and hence molecules again. However, if both surfaces of the zirconia are coated with a metallic electrode, say porous platinum, and the electrodes are connected by a wire, the reaction will continue. The electrons liberated at the high pressure side will traverse the external circuit and reunite with ions arriving at the low pressure side. The force driving the reaction will be the oxygen pressure differential.

The measure of this force is usually taken in thermodynamic terms to be the free energy of the reaction taking place. The free energy of the reaction, ΔG, is given by the equation

$$\Delta G = -nEF$$

where n is the number of electrons transferred during the reaction, F is the Faraday constant and, most importantly, E is the voltage which develops

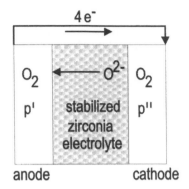

Figure 5.5 The processes taking place in a stabilized zirconia oxygen sensor. The electrolyte separates oxygen gas at two different pressures. Electrons traverse the external circuit from the anode to the cathode, and oxygen ions cross the electrolyte from the cathode to anode. The voltage in the external circuit provides a measure of the oxygen pressure difference across the electrolyte.

between the electrodes as a result of the reaction. So, the end result is that the pressure difference has produced a voltage.

Thermodynamics tells us that the equation for ΔG applies to a much wider range of situations than just oxygen gas at two different pressures. A free energy change and hence a voltage will be produced as long as oxygen is present at two different activities. Now oxygen, which may be part of an oxide, dissolved in a metal or present in the blood, has measurable activities and so can be used as the source of a voltage. The rather wide range of devices which are based on a stabilized zirconia use this voltage as a signal to give information about the difference in the oxygen activity on each side of the stabilized zirconia barrier.

5.4.3 Oxygen sensors

One of the most important applications of calcia-stabilized zirconia is as an *oxygen sensor*. For the situation illustrated in Figure 5.5 the voltage across the zirconia is

$$E = \left(\frac{RT}{4F}\right) \ln\left[\frac{p'_{O_2}}{p''_{O_2}}\right]$$

where R is the gas constant, T the temperature in K, the factor four is the number of electrons required to transform one oxygen molecule into oxide ions and the oxygen pressures are measured in atmospheres. This can be rearranged to give

$$p'_{O_2} = p''_{O_2} \exp\left[\frac{4EF}{RT}\right]$$

An easy way to make an oxygen sensor is to use a stabilized zirconia tube and to coat the inside and outside with porous platinum to which leads are connected. The tube can be used directly as an oxygen meter if p''_{O_2} is a standard pressure, such as 1 atmosphere of oxygen or else the pressure of oxygen in air, which is approximately 0.21 atmosphere. Such a system is utilized to monitor the oxygen content, and thus fuel efficiency, of a car engine. The schematic arrangement is shown in Figure 5.6. The coated zirconia tube is arranged to project into the exhaust stream of the engine. Using air as the standard oxygen pressure, the output voltage of the sensor is directly related to the stoichiometry of the air–fuel mixture. The cell voltage may be used to alter the engine input fuel–air mix automatically so as to optimize engine efficiency. Calibration of the response of the engine as a function of input conditions allows for the construction of a useful reproducible sensor.

The same arrangement can be used to measure the concentration of oxygen in solutions such as liquid metals or blood, and such sensors are in routine use in such diverse areas as medicine and the steel industry. Because the oxygen is dissolved, the voltage measured depends on the activity of the oxygen in the solution. For low concentrations, the activity is equal to the concentration itself, and it is found that

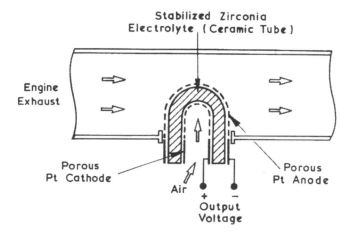

Figure 5.6 A car exhaust sensor using a stabilized zirconia ceramic tube electrolyte. [Redrawn after R.E. Newnham, *Crystallogr. Rev.* **1**, 253 (1988).]

$$E = \left(\frac{RT}{4F}\right) \ln\left[\frac{p_{O_2} \text{ reference}}{O_2 \text{ solution}}\right]$$

If we take p_{O_2} as 1 atmosphere we can simplify the last equation to write

$$E = -\left(\frac{RT}{4F}\right) \ln[O_2 \text{ solution}]$$

These equations consider the oxygen to be present as molecules in solution. If the oxygen exists as atoms the equation is

$$E = -\left(\frac{RT}{2F}\right) \ln[O \text{ solution}]$$

This is because the number of electrons required to transform the oxygen into ions is now two rather than four, as in the case of molecules.

5.4.4   Free energy meters

Because the potential developed across a stabilized zirconia electrolyte is simply related to the free energy of the reactions taking place in the surrounding cell, the material can be used to construct a free energy meter. One of the simplest arrangements that we can envisage is one in which the stabilized zirconia separates oxygen gas at 1 atmosphere pressure and a metal–metal oxide mixture. The voltage measured is directly related to the free energy of formation of the metal oxide. A more useful arrangement, in which a non-stoichiometric oxide is used on one side and a stoichiometric oxide on the other, allows one to measure the free energy of a phase as a function of its composition

We can illustrate this principle using the non-stoichiometric oxide $\sim$FeO as the test material and NiO as the standard, as shown in Figure 5.7. Provided that the metal oxides are mixed with some metal the reactions taking place are

anode reaction : $(1 - x)\text{Fe} + \text{O}^{2-} \longrightarrow \text{Fe}_{1-x}\text{O} + 2e^-$

cathode reaction : $\text{NiO} + 2e^- \longrightarrow \text{Ni} + \text{O}^{2-}$

overall cell reaction : $(1 - x)\text{Fe} + \text{NiO} \longrightarrow \text{Fe}_{1-x}\text{O} + \text{Ni}$

where Fe_{1-x}O represents the non-stoichiometric oxide in contact with Fe metal. Now the voltage measures the difference in the free energies of formation of the oxides and as NiO is a standard

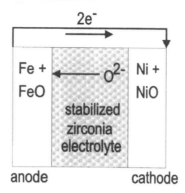

Figure 5.7 The reactions taking place in a stabilized zirconia cell used to measure the free energy of formation of ~FeO. The free energy of NiO is a known standard. The voltage in the external circuit is a measure of the difference in the free energy of NiO and ~FeO.

$$\Delta G^\circ_{FeO} = -2EF + \Delta G^\circ_{NiO}$$

where ΔG°_{FeO} represents the free energy of formation of $Fe_{1-x}O$ and ΔG°_{NiO} is the free energy of formation of NiO.

In this arrangement the voltage will be constant. Now if the iron in the mixture is used up the reaction will be slightly different, as the non-stoichiometric $Fe_{1-x}O$ will still be able to take up oxygen. The reaction will be

$$Fe_{1-x}O + yNiO \longrightarrow Fe_{1-x}O_{1+y} + yNi$$

The voltage measured will now appear to drift as the composition range of the non-stoichiometric oxide is crossed. The magnitude of the drift will be a direct measure of the difference in the free energy of formation of the oxide in contact with Fe, i.e. $Fe_{1-x}O$, with that of the composition produced, $Fe_{1-x}O_{1+y}$.

5.4.5 Oxygen pumps and coulometric titrations

In the arrangements described above, oxygen is transferred from the high pressure (or high free energy) side to the low pressure (or low free energy) side of the zirconia as long as current is flowing in the external circuit. The amount of oxygen transferred can be precisely calculated by an application of Faraday's law of electrolysis. The moles of O^{2-} transferred during cell operation is

$$\text{moles of } O^{2-} = \text{current (A)} \times \frac{\text{time (s)}}{2F(C\,mol^{-1})}$$

This means that we can use the experimental procedure just described to vary the composition of the ~FeO phase in a quite precise way, simply by allowing a known amount of current to flow in the external circuit. This procedure is known as *coulometric titration*.

The process can also be reversed to make an *oxygen pump*. The principle is exactly the same as that used when charging a car battery. If a higher voltage is applied to the electrodes on each side of the stabilized zirconia, and the polarity of the voltage is reversed, oxygen ions will be pumped from the low pressure side to the high pressure side. These pumps have no moving parts or electric motors and are, therefore, of most use in situations where mechanical failure or electrical sparks must be avoided.

5.5   Case study: fuel cells

A battery uses the energy supplied by a spontaneous chemical reaction to provide an electric current. When the supply of reactants on each side of the electrolyte has been exhausted the chemical reaction stops and the battery will no longer serve a useful purpose. *Fuel cells* have been designed to overcome this disadvantage. Broadly speaking the supply of reactants to each side of the cell is continuous and the supply of current that can be drawn is limitless.

There are a number of good reasons for wanting to make a fuel cell, but two of the most important are economy and environmental protection. From the viewpoint of economics, fossil fuels are in ever shorter supply and an alternative source of energy will be necessary sooner or later. Looking at environmental protection, electric motors are non-polluting, both from the point of view of emissions and from the point of view of noise. Battery-driven motors would offer considerable advantages. However, battery production is not always so clean and fuel cell-driven motors would bypass this problem. Moreover, the burning of fossil fuels is at the heart of worries over global warming. Fuel cells would eliminate this worry as well.

Considerations of this nature have driven fuel cell research for a number of years. The ideal reactants for a fuel cell are hydrogen and oxygen. The reaction to form water has a suitable free energy and will yield about 1.2 V. Moreover, the reaction product is benign and oxygen is readily available simply by using air as one component of the cell. Practical cells have had two stumbling blocks, an easy source of hydrogen as a reactant and a suitable electrolyte.

In cells for experimental purposes or when costs are of secondary importance, hydrogen gas and an aqueous solution of KOH have been used as fuel and electrolyte. However, for many purposes a solid electrolyte is to be preferred. If cells are to be used to power cars, for example, a solid electrolyte would prove convenient and safe. Thus, a large number of cells have been constructed using stabilized zirconia as an electrolyte. A schematic diagram of one such cell design is shown in Figure 5.8. In this cell the fuel flows over the anode, which might be porous platinum, for example. The oxygen flows over the cathode, which is also porous platinum. The oxygen gas is oxidized and the stabilized zirconia transports oxygen ions from the high oxygen pressure cathode region to the anode region. The hydrogen fuel reacts with the O^{2-} ions to produce water and electrons which flow to the anode. The cell reactions taking place are

anode reaction: $H_2 (g) + O^{2-} \longrightarrow H_2O + 2e^-$

cathode reaction: $O_2 (g) + 4e^- \longrightarrow 2O^{2-}$

cell reaction: $2H_2 (g) + O_2 (g) \longrightarrow 2H_2O$

For efficient production of power, these cells have to be used at about 700 °C or higher. This is not a problem for automotive use as the cells can be warmed using, at least in part, surplus heat from the engine. The fact that such cells are not used routinely in cars lies in part with the problem of fuel. Despite a great deal of research on non-stoichiometric hydrides such as TiH_x and NbH_x, which are able to store hydrogen gas in a convenient form, no commercially successful portable hydrogen reservoir has yet been perfected. Recent research, in fact, suggests that a conventional hydrocarbon fuel might be used indirectly, by passing it over a catalyst so as to generate hydrogen gas. Although the technology to make fuel cells an economic source of either static or mobile electricity has not quite been achieved, the rewards that will flow from a success will ensure that these cells will eventually be available commercially.

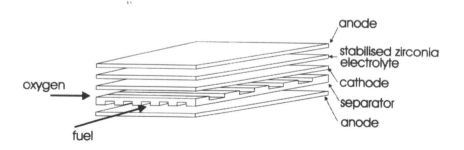

Figure 5.8 Detail of a fuel cell constructed of corrugated sheets of stabilized zirconia. The anode and cathode can be porous platinum, but electrically conducting oxides such as doped $LaMnO_3$ are usually used.

5.6   The β-alumina oxides

5.6.1   High energy density batteries: the problem

An important goal in recent years has to make a battery suitable for use as a power source for vehicles. For this purpose a large cell voltage is required. Because of the connection between voltage and free energy, given by

$$\Delta G = -nEF$$

this is equivalent to finding a reaction associated with a large free energy change.

The reactions between alkali metals and non-metals such as oxygen fall into this group. In terms of energy output per unit weight, these reactions yield about $600\,kW\,h$ per kg of material. In fact the LiI battery, described earlier in this chapter, utilizes this type of reaction. However, the ionic conductivity via the Schottky defects present in the LiI is too low to yield the high currents that are required in vehicles. A fast ion conductor is needed. Suitable materials have been found in a group of oxides related to the compound β-alumina. These have high ionic conductivities, with transport numbers for the alkali metal cations close to 1.0 over a wide range of temperatures (see Figure 5.1) and appear perfect for the job.

5.6.2   The structure of β-alumina

The β-alumina family of phases are all non-stoichiometric compounds with compositions lying somewhere between the limits MA_5O_8 and $MA_{11}O_{17}$, where M represents a monovalent cation, typically Li, Na, K, Rb, Ag or Tl and A represents a trivalent ion, usually Al, Ga or Fe. The parent phase, β-alumina itself, has a nominal composition of $NaAl_{11}O_{17}$.

In fact, both the name and formula of β-alumina are misleading. When β-alumina was first prepared it was thought to be a polymorph of alumina, hence the name. Only later was it discovered that the compound was, in fact, a sodium aluminium oxide. Despite this fact, the name β-alumina was still retained and today is firmly entrenched in the literature.

The general features of the β-alumina structure were clarified as long ago as 1931 and are shown in Figure 5.9. It is seen that the unit cell is composed of two units, the 'spinel blocks' and regions between these blocks which hold them together. The spinel blocks are composed of four oxygen layers in a cubic, close-packed arrangement. In these layers, the Al^{3+} ions occupy octahedral and tetrahedral positions, so the structure resembles a thin slice of the compound spinel, $MgAl_2O_4$, but without the Mg^{2+} ions. So, although these sheets are called spinel blocks, they are neither exactly of the spinel composition, nor blocks, being in fact sheets of unlimited extent in the direction normal to the c-axis. They are held together by a few AlO_4 tetrahedra. This means that the spinel blocks are easily separated, and β-alumina cleaves readily into mica-like foils along these planes. The Na^+ ions reside in this almost empty region between the *spinel* blocks, called the *conduction planes*.

When the structure is determined carefully, problems arise because it is impossible to locate the Na^+ ions with precision. There are three possible Na^+ sites in the conduction planes as shown in Figure 5.10(a). At room temperature, the Na^+ ions are moving continuously between them, as illustrated in Figure 5.10(b). This is just like the situation in AgI. This means that the Na^+ ions behave as a *quasi-liquid* layer throughout the conduction planes. It is now easy to understand why β-alumina is such a good conductor of Na^+ ions. We have almost unimpeded motion in the Na^+ layers and the conductivity is of the same order of magnitude as one would find in a strong solution of a sodium salt in water. This is exactly as desired!

When we compare the situation in β-alumina with that in LiI, we can see that the problem of obtaining a high point defect concentration in the electrolyte has been solved by effectively segregating the 'defects' into Na^+-containing layers, away from the normal *spinel*-like structure which contains only low point defect concentrations. This is indeed a clever structural way for aluminium oxide, which is built up of small cations, to incorporate large cations within its structure without excessive lattice strain.

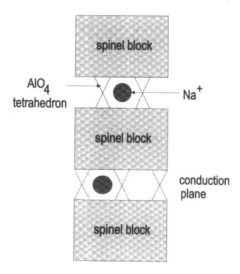

Figure 5.9 The structure of β-alumina shown as a packing of '*spinel* blocks', separated by AlO_4 tetrahedra. The conduction planes contain only Na^+ ions, shown as filled circles, and oxygen ions which form the central apex of the AlO_4 tetrahedra.

The story of the non-stoichiometric structure of β-alumina is not quite finished. While the ideal composition is $NaAl_{11}O_{17}$, the real composition is quite variable and it is found that that the phase always contains an excess of alkali metal. (A typical analysis would yield a composition of $Na_{2.58}Al_{21.81}O_{34}$.) Because crystals of β-alumina contain an excess of Na^+ ions over the idealized formula it is necessary to look for some sort of counter-defects. There are two reasonable possibilities that can be envisaged; the introduction of Al^{3+} vacancies into the *spinel* sheets or else the incorporation of extra oxygen ions into the structure. It is not easy to see how these extra O^{2-} ions can be introduced into the *spinel* sheets as the available unoccupied positions are too small to accommodate them. Therefore, it is necessary to conclude that the extra O^{2-} ions must be present in the same layers as the Na^+ ions.

It has not been easy to obtain experimental evidence to confirm which of these alternative possibilities holds, but on balance it seems that extra oxygen ions enter the Na planes to maintain the charge balance. However, the energy needed to create an Al^{3+} vacancy must be very similar to that

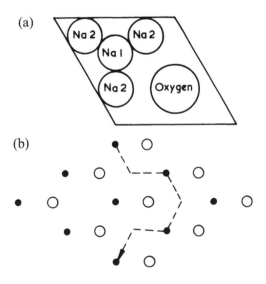

Figure 5.10 (a) Some of the possible Na sites in the conduction plane of sodium β-alumina. The outline marks the extent of the unit cell. The positions labelled Na1 are on average occupied 80% of the time while the sites labelled Na2 are only occupied about 17% of the time. Other sites make up the other 3% occupancy. The oxygen ion forms part of an AlO_4 tetrahedron. (b) The structure of the conduction plane in which the open circles represent O^{2-} ions and the filled circles Na^+ ions in Na1 sites. One of many diffusion paths is shown as a dotted line.

needed to place an oxide ion into the Na$^+$ planes because in some other β-alumina phases Al^{3+} vacancies, or a combination of both vacancies, and extra oxygen occurs.

5.6.3 Other β-alumina related phases

The stacking of the *spinel* blocks in β-alumina is not unique and an alternative arrangement in which the unit cell contains three *spinel* blocks, not two, is possible. This structure is called β''-alumina. Although the structure is different the phase has the same composition and formula as β-alumina. Complex structures which have the same chemical formula but differ in the way in which parts of the structure are stacked on top of one another are often called *polytypes*. The β''-alumina phase is preferred for batteries as it generally shows a higher ionic conductivity for Na$^+$ than the β-form.

The ionic conductivity is due to the alkali metal ions between the *spinel*-like slabs and so is specific to these alkali metal cations only. Thus, if we wish to make a sensor capable of detecting only sodium we could make use of sodium β''-alumina. The mobility of the alkali metal ions has a further consequence. If crystals of sodium β''-alumina are placed in contact with a liquid phase containing another cation, such as molten KCl, then the Na$^+$ will exchange with the K$^+$ and we can make a crystal in which the alkali metal ions are K$^+$ rather than Na$^+$. In this way β''-alumina crystals containing a wide range of monovalent, divalent and lanthanide cations have been prepared. All of these have very similar structures to sodium–β''-alumina but the details of the defect structure vary from one phase to another.

5.6.4 β-alumina batteries: a solution

The ideal properties of the β-alumina oxides were first exploited in a battery by the Ford Motor Company in 1966. As expected, it had an extremely high power density, equal to 1030 W h kg^{-1}. The principle of the battery is shown in Figure 5.11. The reaction chosen was that between sodium and sulphur. The β''-alumina electrolyte, made in the form of a large test-tube, separates molten sodium from molten sulphur, which is contained in a porous carbon felt. The operating temperature of the cell is high, about 300 °C, which is a drawback. However, the cell reaction is extremely energetic and the heat required to maintain the cell at its operating temperature is readily supplied by the cell itself.

The following reactions take place

anode reaction : $2Na \longrightarrow 2Na^+ + 2e^-$

cathode reaction : $2Na^+ + S + 2e^- \longrightarrow Na_2S$

overall cell reaction : $2Na + S \longrightarrow Na_2S$

The anode reaction takes place at the liquid sodium–β''-alumina interface. Here sodium atoms loose an electron and the Na^+ ions formed enter the conduction planes in the electrolyte. The cathode reaction, which occurs at the interface between the β''-alumina and the liquid sulphur, forms sodium polysulphides. The equations given above are representative of this type of reaction.

5.7   The $\overline{Li_xTiS_2}$–Li_3N battery: role reversal

5.7.1 Li_xTiS_2: a non-stoichiometric electrode

In previous sections of this chapter, fast ion conductors have mainly featured as solid electrolytes. Although the electrolyte is a vital component, it is not the only one of importance. The electrode materials also pose severe

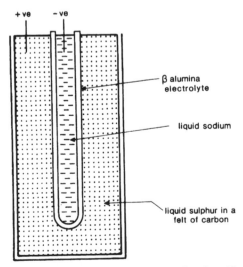

Figure 5.11 Schematic diagram of a sodium–sulphur cell using β''-alumina as a solid electrolyte.

problems which can often be overcome by the use of non-stoichiometric compounds. In this final section, the way in which one material, Li_xTiS_2, is used will be outlined.

In the previous chapter, the structure of non-stoichiometric materials, such as Li_xTiS_2, was described. Li_xTiS_2 is made up of rather dense TiS_2 sheets interleaved with variable amounts of Li. The reaction by which the Li is inserted, called an *intercalation reaction,* is easily *reversible.* This property makes Li_xTiS_2 very suitable for use as electrode material in batteries in which Li^+ ions cross the electrolyte, as it can act as a reservoir for the very reactive Li needed. In general, it is used as the cathode. The reactions which take place when the battery is used (discharged) or recharged are

discharge : $xLi \mid TiS_2 \longrightarrow Li_xTiS_2$

recharge : $Li_xTiS_2 \longrightarrow xLi + TiS_2$

The cell reactions are

anode reaction : $xLi \longrightarrow xLi^+ + xe^-$

cathode reaction : $TiS_2 + xLi^+ + xe^- \longrightarrow Li_xTiS_2$

overall cell reaction : $xLi + TiS_2 \longrightarrow Li_xTiS_2$

which takes place for values of x which are less than 1.

5.7.2   Li₃N: a stoichiometric electrolyte

There are a number of batteries which have been constructed using Li as the anode and Li_xTiS_2 as the cathode. Of most interest in the present context are those which use the material Li_3N as the electrolyte. A diagram of a typical Li_3N battery is shown in Figure 5.12.

How is a stoichiometric compound like Li_3N is able to provide sufficiently high ionic conduction? Structurally, this phase is composed of compact layers of Li and N atoms joined together by Li atoms between the layers as shown in Figure 5.13. There is some uncertainty about the bonding in Li_3N, but it would appear that an ionic model is not too far from the truth. In this case, the layers have a formula of $(Li_2N)^-$ and they are linked by Li^+ ions. Pure Li_3N is a conductor of Li^+ ions. Looking at Figure 5.13, it would be natural to assume that the ionic transport takes place exactly as in β-alumina and that the inter-layer lithium ions are responsible for the conductivity. Quite remarkably, this is false as the phase is strictly stoichiometric. Instead, the conductivity comes about by the formation of Frenkel defects. Some of the Li^+ ions in the hexagonal layers move into interstitial sites in the region between the layers, leaving behind a cation

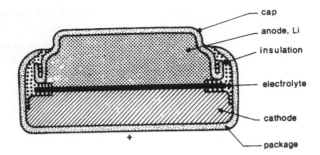

Figure 5.12 Schematic illustration of a Li₃N solid state battery. The electrolyte is stoichiometric Li_3N and the cathode is non-stoichiometric Li_xTiS_2.

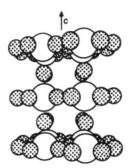

Figure 5.13 The hexagonal structure of Li_3N which contains layers of Li^+ ions, shown as small spheres, and N^{3-} ions, shown as large spheres, connected by bridging Li^+ ions.

vacancy. The energy required to form such a defect is only 0.19×10^{-19} J. Surprisingly, the migrating defects appear to be the Li^+ vacancies and *not* the interstitial Li^+ ions between the layers! The enthalpy of migration of the defects is about 0.19×10^{-19} J, the same as the formation energy of the Frenkel pair. The interacting roles of defects and non-stoichiometry are strangely reversed in this phase!

These cells present the rather unusual feature that a non-stoichiometric compound is used for the cathode and a stoichiometric material for the electrolyte.

5.8   Supplementary reading

There are a number of text books, volumes of conference proceedings and articles devoted to solid state electrolytes which cover the topic in this chapter from the point of view of cells and batteries. A selection of these are:

S. Geller (ed.), *Solid Electrolytes (Topics in Applied Physics)*, Springer Verlag (1977).

P. Hagenmuller and W. van Gool (eds.), *Solid Electrolytes: General Principles, Characterisation, Materials, Applications*, Academic Press (1978).

F.W. Poulsen, N.H. Andersen, K. Clausen, S. Skaarup and O.T. Sørensen (eds.), *Fast Ion and Mixed Conductors*, Risø international Symposium on Metallurgy and Materials Science, Risø National Laboratory, Denmark (1985).

H. Rickert, 'Solid State Electrochemistry', in *Treatise on Solid State Chemistry*, Vol. 4, ed. N.B. Hanny, Plenum (1976).

C.A. Vincent, *Modern Batteries*, Arnold (1984).

K. Funke, in *Progress in Solid State Chemistry*, Vol. 11, eds. J.O. McCaldin and G. Somorjai, Pergamon, Oxford (1976), pp. 345–402.

There are a number of shorter, very readable articles which are worthy of studying, including:

D.J. Fray, Sensors based on solid electrolytes, *J. Mater. Ed.* **9**, 33 (1987).

M.D. Ingram and C.A. Vincent, Solid state ionics, *Chem. Britain* **20**, 235 (1984).

R.E. Newnham, Structure–property relationships in sensors, *Cryst. Rev.* **1**, 253 (1988).

A. Rabenau, Lithium nitride and related materials, *J. Ed. Mod. Mater. Sci. Eng.* **4**, 493 (1982).

N.E.W. de Reca and J.I. Franco, Crystallographic aspects of solid electrolytes, *Crystallogr. Rev.* **2**, 241 (1992).

S. Skaarup, Solid electrolytes, *J. Mater. Ed.* **6**, 667 (1984).

6 Non-stoichiometry and electronic conduction

6.1 Introduction

So far, discussion has been mainly restricted to materials in which the cations took only one valence state. We now add another level to our understanding of the structure and properties of non-stoichiometric compounds by considering materials in which some of the cations can take more than one valence state. Such cations are typically those of the transition metals, but some other atoms, such as Sn or Bi are also of importance.

All of the features of point defect chemistry that have been discussed previously also apply to these compounds. In particular, non-stoichiometry must be accompanied by compensating defects which maintain overall electrical neutrality in the material. Previously these compensating defects were supposed to be vacancies, interstitials or substituted atoms. In materials in which cations can take more than one valence state, electrical neutrality can also be preserved by a change in the charges on the cations. This results in the formation of new sorts of defects which are *electronic* in nature. An understanding of electronic defects will provide a key to the important electrical properties of transition metal compounds and the way in which these properties can be manipulated for our own purposes.

6.2 Non-stoichiometry in pure oxides

6.2.1 Metal excess

An oxide with an approximate formula MO and an experimentally determined metal excess can accommodate non-stoichiometry in two ways. In the absence of any other vacancies or interstitials, these compositional changes cannot be made without introducing electronic defects as well.

Type A materials. In type A materials, anion vacancies are the cause of the metal excess and the oxide will have a real formula MO_{1-x}. In order to keep the crystal neutral we need to introduce two electrons for each oxygen ion moved. A good site for these electrons is a cation. In the present case, the starting oxide will contain only M^{2+} cations. If one electron is associated with one cation, it will change from an M^{2+} ion to an M^{+} ion. This situation is illustrated in Figure 6.1(a).

Type B materials. In type B materials, interstitial cations cause the excess metal and the oxide will have a real formula of $M_{1+x}O$. In order to maintain charge neutrality, each interstitial M^{2+} cation atom must be balanced by two electrons. As before, these will create two M^{+} ions in the structure, as shown in Figure 6.1(b).

In both these materials, non-stoichiometry involves the introduction of extra electrons. If sufficient energy is supplied to make the move from one cation to another, the crystal will be able to conduct electricity. Often light will provide the energy needed to allow the electrons to migrate and the material is said to show *photoconductivity*. When thermal energy alone is sufficient to free the electrons, the electronic conductivity will rise with increasing temperature. Materials behave in this way are called *n-type semiconductors*.

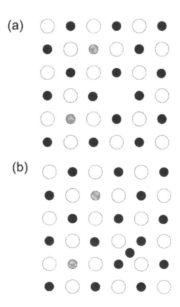

Figure 6.1 Schematic illustration of ways of accommodating changes in stoichiometry in a cation excess oxide of composition close to *MO*. The anions are shown as open circles and the cations are represented as full circles. In (a) an anion vacancy is accompanied by two M^{+} cations, shown lightly shaded. In (b) a cation interstitial is accompanied by two M^{+} ions, which are shown lightly shaded.

6.2.2   Oxygen excess

An oxide, *MO*, with an excess of the non-metal can also accommodate the change in composition in two ways and uses electronic defects to maintain charge neutrality.

Type C materials. In these materials, interstitial O^{2-} anions are the cause of the composition change and will give the oxide a real formula of MO_{1+x}. Figure 6.2(a) shows that to compensate two M^{2+} cations have been converted to two M^{3+} cations.

Type D materials. In these materials, cation vacancies are the reason for the oxygen excess, which will give the oxide a formula of $M_{1-x}O$. This change can be balanced by changing two M^{2+} ions into two M^{3+} ions, as shown in Figure 6.2(b).

Just as we regarded an M^{2+} ion plus an electron as an M^{+} ion, we can regard the M^{3+} ions in type C and D materials as M^{2+} ions plus a trapped *positive hole* or simply a *hole*. If the holes are able to gain enough energy to move from one cation when illuminated, the materials are photoconducting. Thermal energy may also be able to liberate the holes. In both cases, the

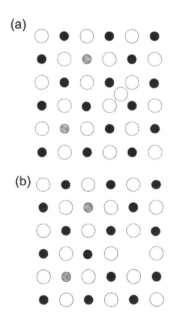

Figure 6.2 Schematic illustration of structural ways of accommodating changes in stoichiometry in anion excess oxides of composition close to *MO*. The anions are shown as open circles and the cations are represented as full circles. In (a) an interstitial anion is accompanied by two M^{3+} cations, shown lightly shaded. In (b), a cation vacancy is accompanied by two M^{3+} cations shown lightly shaded.

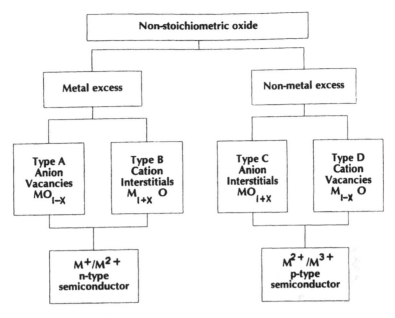

Figure 6.3 A schematic representation of the structural and electronic consequences of non-stoichiometry in oxides of metals with a variable valence.

materials will behave as if the charge was transported by positive particles and they are known as *p-type semiconductors*.

The situation described above is summarized in Figure 6.3. Type B, *n*-type semiconducting materials are well exemplified by ZnO and CdO, while type D, *p*-type semiconducting materials are exemplified by the oxides NiO, Cu_2O, CoO and MnO.

Clearly, in all these materials the number of electronic defects is closely related to the composition. Thus, by controlling stoichiometry, we can control electronic conductivity, a fact of some importance in the search for new electronic materials.

6.3 The effect of impurity atoms

Although the principles just outlined are very important, in practice it is often difficult to alter the composition of a phase to order, as later chapters in this book will make clear. Nevertheless, the importance of electrical properties makes it worthwhile searching for simpler ways of introducing electronic defects into crystals. Fortunately, similar electronic effects can be generated if we can incorporate into the lattice an impurity ion of a different nominal charge or valence to that of the parent atoms. Nowadays the deliberate addition of impurities to cause specific changes in electronic properties of materials lies at the heart of much of the modern electronics industry. These ideas will be illustrated with respect to the semiconductors silicon and germanium, as well as more complex oxide materials.

6.3.1 Impurities in silicon and germanium

The deliberate addition of impurities to Si and Ge so as to modify their electronic conductivity is called *doping*. Both of these elements crystallize with the *diamond* type structure shown in Figure 6.4. In this structure, each atom is surrounded by four others arranged at the corners of a tetrahedron. Each atom has four s^2p^2 outer electrons and each is used up in forming the four sp^3 bonds which connects any one atom to its neighbours.

Consider what will happen if we dope Si with a very small amount of an impurity from the neighbouring group of the periodic table P, As or Sb. The valence electron structure of the impurity atoms is s^2p^3 and after using four electrons to form the four sp^3 bonds, one electron per impurity is left over. These electrons are easily liberated from the impurity atoms by thermal energy and the doped material has become an *n*-type semiconductor. The atoms P, As or Sb in Si or Ge are called *donors* as they *donate* an electron to the crystal.

An analogous situation arises if we dope with elements from the Al, Ga, In subgroup. In this case, the impurities have only three outer electrons in an s^2p^1 configuration which is not sufficient to complete four bonds to the surrounding atoms. One bond is an electron short. It makes life easier if we call the missing electron a positive hole and so each impurity is thought of as introducing one positive hole into the array of bonds within crystal. These impurities are called *acceptors*, because they can be thought of as *accepting* electrons from the otherwise filled bonds. Thermal energy is sufficient to allow these holes to jump from one position to another and these materials are, therefore, *p*-type semiconductors.

6.3.2 Impurities in simple oxides

A well-studied example of the effects of impurity interaction in simple oxides is provided by the $NiO–Li_2O$ system. If colourless Li_2O and green NiO are heated together at high temperatures, the mobile Li^+ ions can easily enter the NiO structure and occupy Ni^{2+} sites. The resulting 'impure' crystal has the formula $Li_xNi_{1-x}O$ where x can take values from 0 to approximately 0.1. This *mixed crystal*, in which the Li has substituted for Ni, is able to form easily because the ionic radii of Ni^{2+} and Li^+ are very similar. The resulting material is black in colour which indicates that something significant has happened.

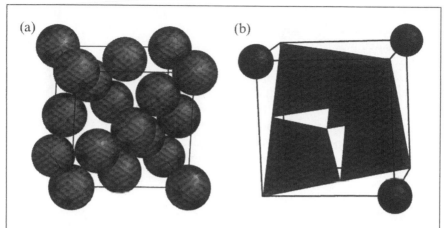

Figure 6.4 The diamond structure type. Perspective view of the cubic unit cell (a) showing atom positions and (b) as carbon-centred tetrahedra.

The diamond structure

The diamond structure is cubic with a unit cell parameter $a = 0.356$ nm. There are eight atoms in the unit cell. Each is connected to its neighbours by four bonds pointing towards the vertices of a tetrahedron. The structure can also be considered to be made up of carbon tetrahedra each with a carbon atom within.

As the Li^+ ions are introduced into the NiO crystals a compensating defect is needed to balance the charge and maintain neutrality, which is in this case the rather uncommon ion Ni^{3+}. Thus, every Li^+ on a Ni^{2+} site in the lattice results in the formation of a Ni^{3+} ion elsewhere.†

For ease of discussion of electronic properties, it is convenient to regard each of these ions as being equivalent to a positive hole located on a Ni^{2+} cation. The reaction, therefore, produces a high concentration of holes located in the nickel oxide. The process of creating electronic defects in a crystal in this way is called *valence induction*. The material is a *p*-type

---

†There is increasing evidence that the real defects in this compound might be O^- rather than Ni^{3+}. That is, the hole sits on an oxygen anion not a nickel cation. Exactly the same is true in the case of many other cationic valences mentioned in this and later chapters and more experimental information is needed to clarify the situation.

semiconductor and as the holes are only weakly bound to the cations, the material shows high conductivity.

It is equally possible to enhance n-type conductivity by suitable doping. For example, consider the consequences of reacting the oxide Ga_2O_3 with ZnO. For small degrees of reaction, the Ga^{3+} is found to substitute for Zn^{2+} and the crystal maintains its overall MO stoichiometry. Neither Ga^{3+} nor Zn^{2+} are ions which take a variable valence and some vacancy compensation mechanism would be expected. However, this does not occur and charge compensation has been found to be electronic in nature. Each Ga^{3+} ion in the lattice is balanced by an electron elsewhere. It is generally believed that these rest on Zn^{2+} ions to generate either uncharged Zn atoms or Zn^+ ions. We can write these reactions in terms of defect chemical equations in the following way

$$Ga_2O_3 \xrightarrow{ZnO} 2Ga_{Zn}^{\cdot} + 2O_O + \frac{1}{2}O_2(g) + 2e'$$

$$Zn_{Zn} + e' \longrightarrow Zn'_{Zn}$$

$$Zn_{Zn} + 2e' \longrightarrow Zn''_{Zn}$$

We will encounter the effects of impurities often in the remainder of this book.

6.3.3   Impurities in complex oxides

Typical examples of complex oxides are provided by the *spinels* AB_2O_4 and the *perovskites* ABO_3. The introduction of impurities into complex oxides will produce similar changes in electronic properties to those described above, if at least one of the cations present is able to change its valence. In fact, such oxides are often chosen for study because the introduction of an impurity cation onto one sub-lattice, say that occupied by A, can change the valence of the cations, B, on the other cation sub-lattice, in a controlled fashion.

A good example is provided by $SrVO_3$. The structure of this phase is of the *perovskite* type, shown in Figure 6.5. The material contains Sr^{2+} ions in the large cage sites and V^{4+} ions in the octahedra. The material behaves like an insulator. If some of the Sr^{2+} ions are replaced by La^{3+} ions, then charge neutrality is maintained by transforming some V^{4+} ions into V^{3+} ions. The V^{3+} ions can be regarded as V^{4+} ions plus a trapped electron, and it is not to difficult to move these electrons from one V ion to another. Thus, although $SrVO_3$ is a poor electronic conductor, $La_xSr_{1-x}VO_3$ is quite a good one. In effect, the insulator has been turned into a metal. We have more to say about electronic conduction in *perovskites* in section 6.7.

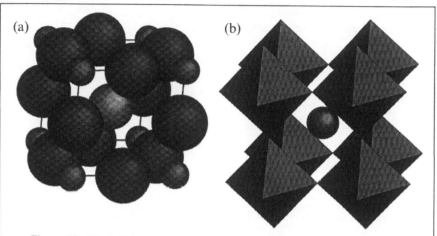

Figure 6.5 The $SrTiO_3$ ideal perovskite structure. (a) The structure shown as an ion array, Ti^{4+} are smallest, Sr^{2+} of medium size and O^{2-} are largest. (b) The structure shown as a packing of corner-shared TiO_6 octahedra with the Sr^{2+} ion located at the cage centre.

The cubic perovskite structure: $SrTiO_3$

The ideal perovskite structure is adopted by the oxide $SrTiO_3$. The unit cell is cubic with a parameter $a = 0.3905$ nm. It consists of an array of corner-sharing TiO_6 octahedra with the large Sr^{2+} ion located at the cell centre. The TiO_6 framework is similar to that in WO_3, shown in Chapter 4.

A large number of oxides of general formula ABO_3, where A is a large cation and B is a medium-sized cation, crystallize with this structure, although in many, slight distortions of the BO_6 octahedra cause the symmetry to change from cubic to tetragonal or orthorhombic.

A rather more subtle way in which 'impurities' can generate electronic defects is found in the Mg–Ti spinels. $MgTi_2O_4$ is a normal *spinel* with a cation distribution $Mg^{2+}[Ti_2^{3+}]O_4$, where the octahedral cations are enclosed in square brackets. However, another *spinel* phase also exists, Mg_2TiO_4. This is an inverse *spinel* with the cation distribution $Mg^{2+}[Mg^{2+}Ti^{4+}]O_4$. What happens if we make a 'solid solution' between these, with a formula $Mg_{2-x}Ti_{1+x}O_4$ in which x varies from 0 to about 0.5? In this material we are replacing octahedral Mg^{2+} by the 'impurity' Ti^{3+}. In the intermediate phases we have a population of Ti^{3+} and Ti^{4+} on the octahedral sites. The Ti^{3+} can be regarded as a Ti^{4+} plus an electron. A comparison with the binary oxides suggests that the material will change from an insulator to an electron conductor and that the conductivity will increase as the value of x increases. This is exactly as found and we have a controlled way of altering the electronic defect population and, therefore, the electronic conductivity of this material.

6.4 Electronic conduction in ionic materials

The materials just described have been discussed in terms of an ionic model. It would, therefore, be useful to set up a theory for electronic conductivity that retained this simple picture. Later in this chapter this model will be linked with another theory of electronic conduction called band theory.

The ionic model of solids treats the normal electrons around each atomic nucleus as being localized at the ions. The electronic conductivity is due to additional electrons or holes created by the defect chemistry of the system. These are also localized or trapped at ions or other defects within the crystal, but not too strongly. They contribute to electronic conductivity by jumping or hopping from one site to another under the influence of an electric field. At a certain time, a localized electron will acquire enough energy to overcome the trapping barrier. It will then move to another site where it becomes relocalized until it gains sufficient energy to make another jump. This is shown in Figure 6.6.

In terms of this picture, a stoichiometric oxide with only one valence state, such as MgO, would be an insulator. This is because if we want to move an electron from one cation to another we must provide energy equal to the

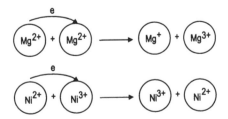

Figure 6.6 In (a) representing a stoichiometric oxide, with cations of fixed valence, such as Mg^{2+}, the electron jump, shown arrowed, requires a prohibitively large energy to create a Mg^+ and a Mg^{3+} ion. For a slightly non-stoichiometric oxide, such as $\sim$NiO (b) already containing a low population of M^{3+} ions, the electron exchange requires only low energy.

further ionization energy of one Mg^{2+} cation and the electron affinity of another Mg^{2+} cation, to result in the hypothetical production of Mg^{3+} and Mg^{+}, as shown in Figure 6.6(a). Such a situation requires so much energy that it is never encountered under normal circumstances and the material will remain an insulator.

On the other hand, if two valence states are available, as in non-stoichiometric $Ni_{1-x}O$, which contains Ni^{2+} and Ni^{3+} ions, an electron jump requires very little energy, because the initial and final state of the crystal are very similar, as shown in Figure 6.6(b). Thus, under normal conditions, electron movement will not be too difficult and the material will be an electronic conductor. Electronic conductivity by a hopping mechanism is, therefore, likely to be restricted mainly to transition metal compounds where alternative valence states are available to cations with little expenditure of energy.

A little thought will show that electron movement by way of discrete jumps is identical to that of atom diffusion discussed in chapter 3. Thus, conduction in materials with hopping-charge carriers is essentially a diffusion process. As treatment of diffusion in this earlier chapter was successful in accounting for many aspects of atom movement, it is worthwhile applying it to the present problem to see exactly where it will lead. In fact, the derivation given in Appendix 6.1 yields the most useful equation

$$\sigma = K(1 - \varphi)\exp\left(-\frac{E}{kT}\right)$$

where K is a constant, φ is the fraction of sites occupied by a mobile electron or hole, $(1-\varphi)$ is the fraction of unoccupied positions that the electron can move to and E is the activation energy for each jump at a temperature of T.

The conductivity, σ, is said to be an *activated process*, that is

$$\sigma \propto \exp\left(-\frac{E}{kT}\right)$$

This means that the conductivity will increase with temperature and hopping materials are often referred to as *hopping semiconductors*.

However, the conductivity will also vary as a function of φ, and this is something new. To illustrate the implication of this fact, consider a non-stoichiometric oxide MO_x in which x can take all values between one and two. Within this composition range suppose three stoichiometric oxides MO, M_2O_3 and MO_2 form. These contain, nominally, M^{2+}, M^{3+} and M^{4+} cations. How exactly will the electronic conductivity vary over the total composition range of the non-stoichiometric phase? The result is given in Figure 6.7.

To understand this answer, suppose that stoichiometric MO_2 is heated in a vacuum so that it loses oxygen. Initially, all cations are in the M^{4+} state and we expect the material to be an insulator. Removal of O^{2-} to the gas phase as oxygen causes electrons to be left in the lattice, which will be localized on cation sites to produce some M^{3+} cations. The oxide now has a few M^{3+} cations in the M^{4+} matrix and thermal energy should allow electrons to hop from M^{3+} to M^{4+}. Thus, the oxide should be an n-type semiconductor. The conductivity increases up to $x = 1.75$, when there are equal numbers of M^{3+} and M^{4+} cations present. As reduction continues, we get to a stage when almost all the ions are now in the M^{3+} state and we have only a few M^{4+} cations left. In this condition, we think of holes hopping from site to site and the material will be a p-type semiconductor. Eventually at $x = 1.5$, all cations will be in the M^{3+} state and we have an insulator, M_2O_3.

We can repeat this argument for the composition range from M_2O_3 to MO. Slight reduction of M_2O_3 will produce a few M^{2+} cations in the M^{3+} matrix, leading to n-type semiconductivity. This would persist in the composition range MO_x between $x = 1.5$ and $x = 1.25$; the conductivity passing through a maximum at the composition $MO_{1.25}$. Further reduction would lead to the situation where we have fewer M^{3+} cations than M^{2+} and we anticipate p-type behaviour in the composition range between $x = 1.25$ and $x = 1.0$. The stoichiometric composition MO should be an insulator.

There are no non-stoichiometric oxides which have a composition range extending all the way from MO_2 to MO, but many which cover a part of this range. These show that the conductivity does vary in the way expected, which gives credence to the idea that hopping conductivity does occur in some phases and that a diffusion model provides a good picture of the process.

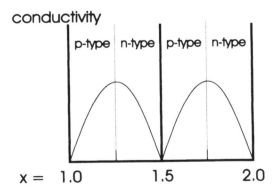

Figure 6.7 Expected variation of the change of conductivity with composition expected for hopping semiconducting oxide MO_x, where x can take values between 1.0 and 2.0.

6.5 Thermoelectric effects

6.5.1 The thermoelectric coefficients

In the materials discussed in this chapter, mobile charge carriers, either electrons or holes, are not too strongly trapped at cation positions and move when sufficient energy is supplied to overcome the activation energy for a jump to the next available site. Thus, we might anticipate that in such materials thermal and electrical effects might be linked. This, in fact, happens and the resulting phenomena are termed *thermoelectric effects*.

The first thermoelectric effect to be discovered was the *Seebeck effect*, which is illustrated in Figure 6.8. When the two ends of a metal or a semiconductor are held at different temperatures, a voltage is produced. This effect is used in the measurement of temperature with a thermocouple. The complementary effect, in which heat is absorbed or generated via a voltage, the *Peltier effect*, is used in refrigeration. Although these effects were discovered quite separately they were shown to be closely related by William Thomson (later Lord Kelvin) who predicted the existence of a third thermoelectric effect now known as the *Thomson effect*. This consists of the appearance of reversible heating or cooling when a current flows along a conductor which has one end at a different temperature to the other.

All three thermoelectric effects are *bulk effects* and the magnitude of the effects produced can be characterized by the materials parameters α, the *absolute Seebeck coefficient*, π, the *absolute Peltier coefficient* and τ, the *Thomson coefficient*.†

Although the three effects are interrelated, the Seebeck effect and the Peltier effect are the most important and are exploited in a wide range of commercial devices. The relationship between these two coefficients is

$$\pi = \alpha T$$

where T is the temperature in K. Thermoelectric parameters are very useful as they give information about the type of mobile charge carriers in the sample. The following section illustrates this by reference to the Seebeck effect.

---

†The terminology and the description of these coefficients is confused in the literature because the effects themselves are usually described with respect to a junction between two different conductors, which introduces unnecessary complications. Values of the Seebeck and Peltier coefficients, derived from such experimental arrangements, are relative values, which, to first approximation, represent the difference between the absolute values for the pair of materials which form the junction. In our discussion we are referring only to absolute values, which are properties of a single component and not a pair of materials. The sources listed in Section 6.7 will give further information on this knotty problem.

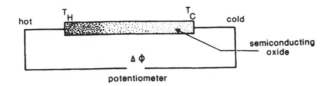

Figure 6.8 The Seebeck effect. The sample, which is typically an oxide such as NiO, is placed in a temperature gradient so that the temperature varies from one end, which is at T_H, to the other at T_C. This results in a potential difference of $\Delta\phi$, between the ends when equilibrium is reached.

6.5.2   The Seebeck coefficient and defect type

The Seebeck coefficient of a material is defined as the ratio of the electric potential produced, measured in volts and when no current flows, to the temperature difference present across a material. Thus, referring to Figure 6.9

$$\Delta\Phi = \Phi_H - \Phi_C = \pm\alpha(T_H - T_C) = \pm\alpha\Delta T$$

$$\alpha = \frac{\Delta\Phi}{\Delta T}$$

where Φ_H and Φ_C are the potentials and T_H and T_C are the temperatures at the hot end and the cold end of the sample, respectively. The units of α are volts per degree. The Seebeck coefficient for metals is of the order of $10\,\mu\text{V}\,\text{K}^{-1}$ and for semiconductors is about $100–300\,\mu\text{V}\,\text{K}^{-1}$.

We can gain an intuitive idea of the cause of the Seebeck effect by assuming that the electrons or holes behave rather like gas atoms. In this case, the charge carriers in the hot region will have a higher kinetic energy, and hence a higher velocity, than those in the cold region. This means that

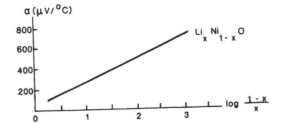

Figure 6.9 Variation of the Seebeck coefficient with defect concentration in $\text{Li}_x\text{Ni}_{1-x}\text{O}$ crystals. In this representation the number of defects decreases from left to right along the x-axis.

the net velocity of the charge carriers at the hot end moving towards the cold end will be higher than the net velocity of the charge carriers at the cold end moving towards the hot end. In this situation more carriers will flow from the hot end towards the cold end than vice versa. This will cause a voltage to build up between the hot and cold ends of the sample. Eventually equilibrium will be established and the potential so set up is called the *Seebeck voltage*. In the case of materials which have mobile electrons, for example *n*-type semiconductors or metals, the colder end of the rod will be negative with respect to the hotter end because of a build up of negative carriers in this region. In the case where the mobile charge carriers are positive holes, for example *p*-type semiconductors, the colder end of the rod will be positive with respect to the hotter end. Thus, a measurement of the sign of the Seebeck coefficient will show whether the material is *n*-type or *p*-type. For example, the non-stoichiometric forms of NiO, CoO and FeO all show positive values for α, indicating that conduction is by way of holes whereas non-stoichiometric ZnO has a negative value of α, indicating electron mobility.

The picture just drawn of mobile particles moving under the influence of temperature is accounted for in thermodynamics by the entropy contribution. It is quite easy to show that the Seebeck coefficient is actually a measurement of the entropy of the charge carriers. The derivation is given in Appendix 6.2. A very simple relationship is found. For electrons

$$\alpha = -\frac{s}{e}$$

and for holes

$$\alpha = \frac{s}{e}$$

where s is the entropy and e is the charge on a mobile carrier. The units of entropy are joules per degree ($J\,K^{-1}$) per particle and the units of the electron charge will be coulombs (C) per particle. Thus, the units of the Seebeck coefficient will be

$$(J\,K^{-1}) \text{ per particle}/C \text{ per particle} = J/CK = V/K$$

because

$$J/C = V$$

so that the units of the Seebeck coefficient will be volts per degree, as expected.

6.5.3   The Seebeck coefficient and defect concentrations

The equations in Appendix 6.2 apply to all materials with mobile charge carriers. However, we can apply these equations to hopping materials and derive a fairly simple relationship between the magnitude of the Seebeck coefficient and the number of defects present in the material. An important result is that the fewer defects present, the larger α becomes. This is particularly useful when we wish to investigate crystals which show only very small departures from stoichiometry as these are often difficult to study by other means.

The derivation set out in Appendix 6.3 shows

$$\alpha = \pm \frac{k}{e} \left[\ln\left(\frac{n_o}{n_d}\right) + A \right]$$

where the positive version applies to p-type materials and the negative expression to n-type materials. In this equation, A is a constant, n_o is the number sites for the mobile electrons or holes to occupy, normally the number of cation sites, and n_d is the number of electrons or holes. The number of electrons or holes is equivalent to the number of defects present and so the value of the Seebeck coefficient will be largest for lowest defect populations.

Example 6.1

Some data for the material $Li_xNi_{1-x}O$ are presented in Figure 6.9. Each Li^+ substitutes for an Ni^{2+} ion in the structure and this results in the formation of one Ni^{3+} ion per Li^+. The number of mobile holes which appear in the structure is equal to the number of Ni^{3+} ions and hence to the number of Li^+ ions incorporated. Therefore, n_d is equal to x. The term n_o is the number of sites that the holes can jump to. This is equal to the number of unchanged Ni^{2+} cations, which is equal to $(1-x)$. The Seebeck coefficient of $Li_xNi_{1-x}O$ is given by

$$\alpha = + \frac{k}{e} \left[\ln\left(\frac{1-x}{x}\right) + A \right]$$

This is a straight-line equation and a graph of α versus $\ln[1 - x/x]$ should be a straight line which increases as the log term increases, i.e. as x decreases. The slope of the graph should be positive for holes, and of a value k/e. This is in good agreement with the data shown in the figure.

It is also possible to use the data to see how well the theory fits numerically. As an example, $\log[(1 - x)/x] = 1$ which yields $x = 0.0909$ and $(1-x) = 0.9091$. The term k/e is given by

$$k/e = \frac{1.381 \times 10^{-23} \, \text{J K}^{-1}}{1.602 \times 10^{-19} \, \text{C}} = 86 \, \mu\text{V K}^{-1}$$

(Remember that J/C = V.) Inserting these values we find

$$\alpha = +86\left[\ln\left(\frac{0.9091}{0.0909}\right) + A\right] = [198 + A]\frac{\mu\text{V}}{\text{K}}$$

The value of α given in the figure is about $250 \, \mu V K^{-1}$. It therefore appears that the theory is quite good and gives reasonable values for the Seebeck coefficient.

6.5.4   The Seebeck coefficient and stoichiometry

Because the value α depends on the number of defects present it should vary systematically with the composition. It is interesting to sketch out, in a qualitative way, this variation for a non-stoichiometric phase. As an example, let us consider a non-stoichiometric oxide MO_2 which is fairly readily reduced to form MO_{2-x}, and which passes through the phases M_2O_3 and MO during the course of this reduction.

The sequence of events which occurs during the reduction was described earlier. Reference to this discussion shows that initial reduction will populate our MO_{2-x} crystal with a few M^{3+} ions which will give rise to n-type semiconduction. The value of α will, therefore, be large and negative, as shown on the far right of Figure 6.10. This value will fall as the number of defects increases, in accordance with our earlier analysis and the curve approaches zero in the figure. Turning to M_2O_3, a slight degree of oxidation will introduce into the M_2O_{3+x} phase a small number of M^{4+} ions in a matrix of M^{3+} ions. This will lead to p-type semiconductivity and a large positive value for α, shown in the centre part of Figure 6.10. Continued oxidation will cause this value to fall as the number of M^{4+} centres increases. This leads to the variation of α over the complete composition range from MO_2 to M_2O_3 shown in the right-hand part of Figure 6.10, where the region between the high α and low α regions has been extrapolated from the values near to the end compositions. Most significantly, there is a change from n-type to p-type behaviour as the composition passes through a composition of $MO_{1.75}$.

A similar situation will hold as we span the composition range between M_2O_3 and MO, with α changing from positive to negative at a composition of $MO_{1.25}$. This composition range is, therefore, identical to that just described, and is shown on the left-hand side of Figure 6.10. As mentioned when discussing conductivity, no oxides are known which span this whole composition range. However, Figure 6.10 shows that the value of α will

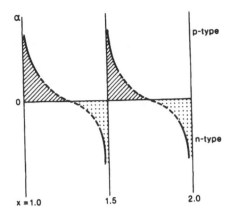

Figure 6.10 Expected variation of the Seebeck coefficient, α, with composition for a non-stoichiometric oxide MO_x, where x can take values between 1.0 and 2.0.

change drastically from large and positive to large and negative as we pass through the stoichiometric position at M_2O_3. Because the discussion leading to this conclusion can be applied to any stoichiometric composition, be it MO, M_2O_3 or MO_2 the effect is easily observable and has been confirmed for a number of non-stoichiometric oxides.

6.6   Band theory

6.6.1   Energy bands

Everything that we have said so far in this chapter has lead to the conclusion that electronic conduction is simply an extension of diffusion theory. In terms of jargon, the electrons or holes are said to be localized and we are really using an ionic model for the conduction process. While this is certainly reasonable for the metal oxides we have described, there are large numbers of materials known for which this description does not seem to be valid. Pure metals and most alloys come into this category, together with many sulphides and some oxides. In order to understand many aspects of defect crystal chemistry we must now consider these materials.

The theoretical approach which has been pre-eminent in describing the electronic properties of metals and alloys is called *band theory*. In band theory, the electrons responsible for conduction are not linked to any particular atom. They can move easily throughout the crystal and are said to be free, or very nearly so. To put this another way, the wave functions of these electrons are considered to extend throughout the whole of the crystal and are said to be delocalized. In this section, a brief and qualitative account of the theory is given and it is shown how semiconducting transition metal oxides, such as NiO or FeO, and impurities in pure materials are described in terms of band theory.

We can summarize the principal results of the theory by stating that the outer electrons in a solid, that is, the electrons which are of greatest importance from the point of view of electronic properties, occupy bands of allowed energies. The way this comes about is easy to understand. In an isolated atom, the electrons occupy a ladder of sharp energy levels which are filled up from the lowest energy to the highest in order. As far as electronic conductivity is concerned only the topmost energy level is of importance. This can be completely filled (with two electrons) or be partly filled (with one). In Figure 5.11(a), the outermost energy level of an isolated atom is shown. If another atom approaches the first the outer electron clouds will interact and the result is that the single energy level will split up into two, one at a higher energy and one at a lower energy, as shown in Figure 6.11(b). The amount of splitting will depend on the closeness of the interacting atoms. The closer they approach, the wider will be the separation of the upper and lower levels. The electrons from both atoms will be placed into the two levels, starting with that at lowest energy. The amount of filling of the energy levels will depend on the number of electrons available.

This process can be continued indefinitely. As each atom is added to the cluster it adds its energy level to the collection and the number of energy levels in the high energy and low energy groups gradually increase. At the same time the spacing between the energy levels in each group decreases. Both of these features are shown in Figure 6.11(c). All of the available electrons are allocated to these energy levels starting at the lowest and filling up towards the highest, exactly as before. Ultimately, when a large number of atoms are brought together to form a crystal, the energy levels in both the high energy and low energy groups are very close indeed. They are now called energy bands and are shown shaded in Figure 6.11(d). The electrons, as before, are poured into the bands to occupy them from the lowest energy level upwards.

The general features shown in Figure 6.11 depend on the degree of interaction of the electron energy levels. This effect is illustrated in Figure 6.12. If the interaction is large, typically when the atoms are close together, the average separation of the upper and lower bands is large but they are also very broad and may overlap. When the interaction is less, as occurs when the atoms involved are further apart, the separation is small, as is the width of each band.

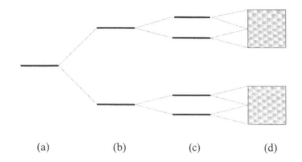

(a)   (b)   (c)   (d)

Figure 6.11 The development of energy bands. (a) A single outer electron orbital on an isolated atom. (b) Two atoms yield two energy levels. (c) Four atoms yield four energy levels. (d) Bands of energy are produced when large numbers of atoms aggregate.

6.6.2 Insulators, semiconductors and metals

The fundamental division of materials, where electrical properties are considered, is into metals, insulators and semiconductors. An *insulator* is a material which normally shows no electrical conductivity. Metals and semiconductors were originally classified more or less in terms of the magnitude of the measured electrical conductivity. However, this turned out to be less than satisfactory and now a better definition is to include in metals

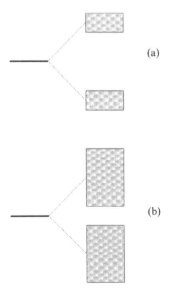

(a)

(b)

Figure 6.12 (a) The band geometry when atoms are rather well separated and have rather small interactions. (b) The band geometry when atoms are rather close and have strong interactions.

those materials for which the electrical conductivity falls as the temperature increases. Semiconductors show an increase in electrical conductivity as the temperature increases. The differences between these groups is explained by band theory in the following way.

Insulators have the upper band completely empty and the lower band completely filled by electrons. The energy gap between the top of the filled band and the bottom of empty band is quite large, as can be seen in Figure 6.13(a). The filled band of energies is called the *valence band* and the empty band is called the *conduction band*. The energy difference between the top of the valence band and the bottom of the conduction band is called the *band gap*. Now when a material conducts electricity, the electrons pick up some energy and are transferred to slightly higher energy levels. If the electron-containing band is full, conductivity cannot occur because there are no slightly higher energy levels available.†

Intrinsic semiconductors have a similar band picture to insulators, except that the separation of the empty and filled bands is small, as in Figure 6.13(b). How small is small? The band gap must be such that some electrons will be transferred from the top of the valence band to the bottom of the conduction band at room temperature. The electrons in the conduction band will have plenty of slightly higher energy levels available and when a voltage is imposed on the material they can take up some energy and this leads to some degree of conductivity. At 0 K these materials will be insulators because no electrons will cross from the valence band to the conduction band. The magnitude of their electrical conductivity will increase as the temperature increases because more electrons will gain sufficient energy to cross the band gap.

Rather surprisingly, an equal contribution to the conductivity will come from positive charge carriers equal in number to the electrons promoted into the conduction band. These are illustrated in Figure 6.13(b) as 'vacancies' in the valence band. We have already met with these before as *positive holes*. Each time an electron is removed from the full valence band to the conduction band two mobile charge carriers are created, an electron and a hole.

Degenerate semiconductors are similar to extrinsic semiconductors but the band gap is similar to, or less than, the thermal energy. In these cases the number of charge carriers in each band becomes very high.

Extrinsic semiconductors contain an appreciable number of foreign atoms which have been added intentionally as dopants. At very low temperatures these may have no effect on the electronic properties of the material but as

†Naturally, if an enormous voltage is applied, as in a thunderstorm, the electrons are given so much energy that they are ripped from the valence band and can transfer to the conduction band. In these conditions the insulator is said to break down.

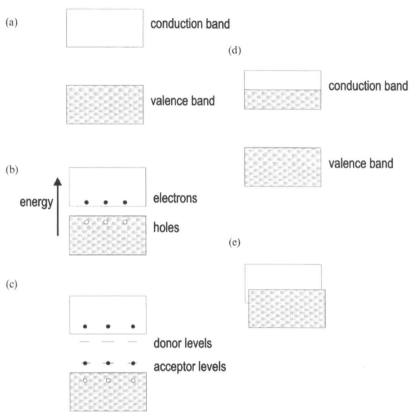

Figure 6.13   Schematic representation of the energy band scheme for (a) insulators; (b) intrinsic semiconductors; (c) extrinsic (doped) semiconductors; (d) metals; and (e) semi-metals.

the temperature rises they can influence the behaviour in two very different ways. They can act as donors, donating electrons to the conduction band, or as acceptors, accepting electrons from the valence band, which is equivalent to donating holes to the valence band. Donors and acceptors are often represented by lines drawn at a distance from the band equal to the energy required for the electron transfer to take place. This is shown in Figure 6.13(c). When donors are the main impurities present in the crystals, the conduction is mainly by way of electrons and the material is called an *n*-type semiconductor. Similarly, if acceptors are the major impurities present, conduction is mainly by way of holes and the material is called a *p*-type semiconductor. This behaviour marks a difference between intrinsic semiconductors and extrinsic semiconductors. In intrinsic semiconductors, electrons and holes are present. In extrinsic semiconduction, the conductivity is due to either one or the other.

If the donors and acceptors are present in equal numbers, the material is said to be a *compensated semiconductor*. At 0 K these materials are insulators, and it is difficult in practice to distinguish between compensated and intrinsic semiconductors. When all of the impurities are fully ionized, so that either all the donor levels have lost an electron or all the acceptor levels have gained an electron, the *exhaustion range* has been reached.

Metals are defined as materials in which the uppermost energy band is only partly filled as shown in Figure 6.13(d). The uppermost energy level filled is called the *Fermi energy* or the *Fermi level*. Conduction can take place because of the easy availability of empty energy levels just above the Fermi energy. In a crystalline metal, the Fermi level possesses a complex shape and is called the *Fermi surface*.

Semi-metals show metallic conductivity due to the overlap of a filled and an empty band, as shown in Figure 6.13(e). In this case, electrons spill over from the filled band into the bottom of the empty band until the Fermi surface intersects both sets of bands. In semi-metals holes and electrons coexist even at 0 K.

6.6.3   Point defects and energy bands

The presence of point defects is represented on band diagrams in a similar way to that shown in Figure 6.13(c). Thus, defects are treated as donors which donate electrons to the conduction band to produce *n*-type semiconductors, or acceptors, which create holes in the valence band to produce *p*-type semiconductors.

Consider the case of interstitial metal atoms added to an insulating ionic oxide. These will tend to act as donors because metal atoms tend to ionize by losing electrons. They are represented by donor levels shown below the conduction band in Figure 6.14(a). Note that the energy of the ionized donor is now lower than that of the neutral interstitial by the same amount as required to move the electron into the conduction band. The material is now an *n*-type semiconductor.

In the case of interstitial non-metal atoms, these will act as acceptors because anion formation involves taking up extra electrons. These defects are represented as acceptor levels just above the top of the valence band, as shown in Figure 6.14(b). They can ionize by taking electrons from the valence band, creating holes in the process, as shown above. Once again, the energy of the neutral acceptor atom is at a different level to that of the ionized acceptor. These materials are *p*-type semiconductors because of the presence of holes in the valence band. The electrons on the ionized anions do not contribute to the conductivity.

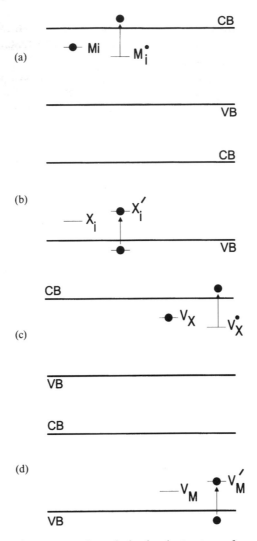

Figure 6.14 Schematic representation of the band structure of an insulating material containing point defects. Each defect is associated with an energy level, drawn as a short line. In (a) the defects are neutral and ionized cation interstitials; in (b) the defects are neutral and ionized anion interstitials; in (c) the defects are neutral and ionized anion vacancies; and in (d) the defects are neutral and ionized cation vacancies. CB indicates the conduction band and VB the valence band.

The same sort of considerations will apply to vacancies. For instance, an anion vacancy gives rise to a set of donor levels just below the lower edge of the conduction band. If the vacancy is created by removing a neutral non-metal atom from the crystal, the electrons that were on the anion are transferred to the conduction band to produce an *n*-type semiconductor, as in Figure 6.14(c). As before, the energy of neutral and ionized vacancies are slightly different.

A cation vacancy will be opposite to this in behaviour. Hence the removal of a neutral metal atom from a material will involve removal of a cation plus the correct number of electrons which are taken from the valence band. Cation vacancies will, therefore, be represented as acceptor levels situated near to the valence band together with an equivalent number of holes in the band as shown in Figure 6.14(d). These materials are *p*-type semiconductors.

To illustrate these features, zinc interstitials in ZnO will be represented by the scheme shown in Figure 6.14(a) as neutral or ionized donor levels. Similarly, cation vacancies in NiO will be represented as neutral or ionized acceptor levels as shown in Figure 6.14(d). The amount of ionization of donor and acceptor levels, and hence their influence on electric properties, will depend on how close the energy is to the nearest band. Hence some impurities produce *deep energy levels*, which have little effect on properties, while others give rise to *shallow levels*, which have a larger effect.

6.7 Band conduction and hopping conduction

The characterization of the defects present in semiconducting or metallic non-stoichiometric compounds is often approached using electrical conductivity measurements. Because of this it will be useful to contrast the equations derived for hopping materials with those best described in terms of band theory. For example, the type of conductivity behaviour shown in Figure 6.8 is quite different from that expected for an alloy of two metals which is well described in band theory terms. In this case, the conductivity is frequently a linear function of the composition and certainly does not tend to zero at the end compositions.

For a typical band theory metal, conducting by electrons, the conductivity, σ, is given by:

$$\sigma = ne\mu$$

where n is the number of mobile electrons in the metal, which is more or less constant over small temperature ranges, and μ is the mobility of the electrons. The mobility of the electrons depends on their mass and the number of times an electron collides with another electron or some other obstacle in the crystal. It can be expressed as

$$\mu = \frac{e\tau}{m^*}$$

where m^* is the *effective mass* of the electron in the crystal. The number of collisions is included as τ, the mean lifetime between electron collisions. At normal temperatures μ is inversely proportional to a low power of temperature, typically

$$\mu \propto \frac{1}{T^{3/2}}$$

For a metal, therefore, σ decreases with temperature

$$\sigma \propto \frac{ne}{T^{3/2}}$$

In the case of semiconductors the equation for the conductivity is

$$\sigma = ne\mu$$

for electrons and

$$\sigma = pe\mu$$

for holes, where n, the number of mobile electrons, is replaced by the number of mobile holes, p. The mobility of the electrons or holes is dependent on temperature, just as in a metal. However, in semiconductors we cannot consider the number of charge carriers, n or p, to be constant. In an intrinsic semiconductor, n or p increases with temperature in the following way:

$$n = p = n_o T^{3/2} \exp\left[-\frac{E_g}{2kT}\right]$$

where E_g is the band gap and n_o is a constant. Very similar equations hold for doped semiconductors, the main difference being that E_g needs to be replaced by the energy required to create an electron in the conduction band or a hole in the valence band. Thus, we find that the conductivity increases with temperature in a semiconductor because of the increase in the value of n or p. So for a semiconductor we can generally write:

$$\sigma \propto \exp\left(-\frac{E}{kT}\right)$$

where E represents the appropriate energy term for electrons or holes.

A comparison of the relevant equations for metals, band theory semiconductors and hopping semiconductors is given in Table 6.1. These equations can be used in a diagnostic fashion to separate one material type from another. Thus, if we find that the electronic conductivity increases with temperature we either have a band-like semiconductor or a hopping semiconductor, but not a metal. If, in the same material, the mobility increases with temperature we must have a hopping semiconductor.

Table 6.1   Electrical conductivity and mobility of charge carriers in metals, band-like semiconductors and hopping semiconductors

	Metal	Band-like semiconductor	Hopping semiconductor
Conductivity, σ	$\propto T^{-m}$ falls slightly with T	$\propto \exp[-E/kT]$ increases with T	$\propto \exp[-E/kT]$ increases with T
Mobility, μ	$\propto T^{-m}$ falls slightly with T	$\propto T^{-m}$ falls slightly with T	$\propto \exp[-E/kT]$ increases with T

6.8   Case study: turning an insulator into a metal

Bearing in mind the wide range of electrical behaviour described above, it is not especially surprising to find that a number of materials exist which transform from metallic behaviour to semiconducting or insulating behaviour or vice versa. Such transitions can be induced by either temperature, pressure or a change in crystal structure. We will not discuss these possibilities here but conclude this chapter by looking at the way in which it is possible to bring this change about simply by utilizing non-stoichiometry!

Let us first look at an example in which the role of stoichiometry is disguised. This is provided by $LaCoO_3$, which has the *perovskite* structure. At temperatures close to absolute zero the material is an insulator. It is made up of La^{3+} and Co^{3+} ions and so would be expected to be an ionic insulator. As the temperature increases to approximately $110\,K$, electron–hole pairs start to be generated. In band theory terms, some electrons are thermally excited from the valence band to the conduction band. In chemical terms, it is possible to regard these electrons as originating in a disproportionation reaction in which electrons on some of the Co^{3+} ions are transferred to neighbouring Co^{3+} ions. This will produce Co^{2+} and Co^{4+} ions in the following way

$$2Co^{3+} \longrightarrow Co^{2+} + Co^{4+}$$

We have the simultaneous generation of two sorts of defect at once, Co^{2+} and Co^{4+} ions, distributed on the Co^{3+} sub-lattice.

The Co^{2+} ions can be thought of Co^{3+} plus a trapped electron while the Co^{4+} ions can be regarded as Co^{3+} plus a trapped hole, i.e. the last equation can be rewritten as

$$2Co_{Co} \longrightarrow Co'_{Co} + Co^{\bullet}_{Co}$$

The Seebeck coefficient of this compound is found to be large and positive.

This must mean that the holes are mobile but the electrons are trapped. This is rather a surprise and tells us that Co^{2+} is stable but Co^{4+} is not, at least in $LaCoO_3$ at temperatures above 110 K!

As the temperature increases to 350 K the degree of disproportionation increases steadily and the number of mobile holes increases. The Seebeck coefficient falls, as expected, but remains positive.

Interestingly, in the temperature regime between approximately 350 and 650 K another change occurs. The semiconducting behaviour gradually changes to metallic behaviour. This comes about because the band gap gradually decreases to zero as the proportion of large Co^{2+} ions in the structure increases. (Remember that the band separation is a function of the electron interactions and because Co^{2+} is much bigger than Co^{3+}, the interactions increase as the degree of disproportionation increases. This makes the conduction and valence bands widen and eventually overlap.) Above about 650 K overlap is complete and the material is metallic.

Can this transition be brought about in a more controlled way using valence induction? The answer is yes. If we make a series of compounds $La_{1-x}Sr_xCoO_3$ we find a transition to the metallic state. When $x = 0$, the material $LaCoO_3$ has a rather high resistivity, as suggested above. When Sr^{2+} is introduced into the material it substitutes for the La^{3+} and occupies La sites. In order to maintain charge neutrality each Sr^{2+} cation introduced into the crystal causes a Co^{3+} cation to change to a Co^{4+} cation. In the composition range up to $x = 0.15$ the material is a p-type semiconductor, as expected. It is possible to say with confidence that the Co^{4+} ions can be regarded as Co^{3+} plus a trapped hole which is able to migrate in the applied electric field. The p-type semiconductivity increases steadily as the value of x increases. In the region between x values of 0.2–0.3 the material changes into a metal, and for a value of $x = 0.3$ the material is completely metallic with a conductivity which decreases as the temperature increases. This is quite opposite to that found in the p-type semiconductor, of course.

How can this be explained? We do not have large Co^{2+} ions to alter the band gap. Instead it is the number of holes present that changes the system. When only one or two holes are present, they will be quite strongly attracted to any of the Co^{3+} cations nearby. As more and more holes form in the crystal, they spread throughout the crystal and form a form a positive cloud draped around the Co^{3+} sub-lattice. This is called a *screening potential* because it hides the Co^{3+} ions from any new holes formed. With only a few holes the screening is not very effective but it does cause the hole trapping energy to decrease. Using Sr^{2+} substitution, though, it is possible to introduce far more Co^{4+} than occurred naturally in undoped $LaCoO_3$ and although the screening also occurs in the undoped material it never reaches significant levels. In $La_{1-x}Sr_xCoO_3$ the shielding has become so effective at

$x = 0.3$ that the trapping energy has been reduced to zero. At this stage the holes become free. No hopping energy is required and the material shows metallic behaviour.

Although an insulator has been turned into a metal in both these examples the transition was caused by two quite different microscopic effects, both of which involved the same defect, Co^{4+} ions in a Co^{3+} sub-lattice. The perovskite system contains many more surprises for us. Some of these will be explored in chapter 10.

6.9   Supplementary reading

There is no compact source of information on the material covered in this chapter, but related aspects, particularly for oxides, will be found in:

M.S. Seltzer and R.J. Jaffee (eds.), *Defects and Transport in Oxides*, Plenum, New York (1975).
C.N.R. Rao (ed.), *The Chemistry of the Solid State*, Marcel Decker, New York (1974).
P. Kofstad, *Nonstoichiometry. Diffusion and Electrical Conduction in Binary Metal Oxides*, Wiley–Interscience, New York (1972).
J.B. Goodenough, *Progress in Solid State Chemistry*, Vol. 5, ed. H. Reiss, Pergamon, Oxford (1972).

The band theory of materials is covered in an approachable way in:

P.A. Cox, *The Electronic Structure and Chemistry of Solids*, Oxford University Press, Oxford (1987).

A very readable review article covering electron transport of in solids is:

D. Adler, *Treatise on Solid State Chemistry*, Vol. 2, ed. N.B. Hannay, Plenum, New York (1975).

Electrical conductivity in oxides is discussed clearly with self-test questions by:

O. Johansen and P. Kofstad, *J. Mater. Ed.* **7**, 909 (1985).

Metal–insulator transitions are treated in depth in:

N.F. Mott, *Metal–Insulator Transitions*, Taylor and Francis, London (1974).

Appendix 6.1
Hopping conductivity

Referring to chapter 3, for a random diffusion process the relationship between ionic conductivity, σ, and self-diffusion coefficient, D, is given by the Einstein relation

$$\sigma = \frac{ne^2 D}{kT}$$

where n was the number of ions that could diffuse per unit volume. As the hopping process will be identical to the diffusion process, this equation will also apply to electron or hole hopping, provided that it takes place by a random series of jumps. In this case, n will be the number of electrons or holes that can move per unit volume of crystal.

The strategy that is now employed is to derive a theoretical expression for the diffusion coefficient, following the route laid down in chapter 3, and use this to obtain a relationship between the conductivity, σ, and other atomic parameters. We, therefore, start with the expression for the number of successful jumps made by a hopping electron, which is

$$q = \nu \exp\left[-\frac{E}{kT}\right]$$

where ν is the attempt frequency for a hop, q is the probability that a jump along the field direction will be successful and E is the activation energy for the hop.

Now such an expression is valid if each possible jump is to an available site. In the case under discussion the electron cannot just jump anywhere. If we consider our example of $Ni_{1-x}O$ an electron can jump from Ni^{2+} to Ni^{3+} but not from one Ni^{2+} to another or from one Ni^{3+} to another. To allow for this, we can designate the number of sites which are occupied by mobile charges, either e' or h', by φ, which is expressed as a fraction of the total sites which the mobile charge carriers can occupy. Thus, the fraction of available unoccupied sites is $(1-\varphi)$. With this proviso, we can now rewrite the last equation in the correct form for us, so that the probability of a successful jump will be given by

$$q = (1 - \varphi)\nu \exp\left[-\frac{E}{kT}\right]$$

We now proceed to calculate the diffusion coefficient of these moving charges, following the procedure in chapter 3. The steps in the argument will, therefore, be only briefly outlined. In the electric field responsible for the electronic conductivity, a gradient of mobile charge carriers will exist. If there is a density of na carriers on the plane at position x, per unit volume, and a density of $[N + a(dn/dx)]a$ and $[n - a(dn/dx)]a$ on the two adjacent planes at time t, we can write

$$\frac{dn}{dt} = \frac{1}{2}q(1 - \varphi)a^2\frac{dn}{dx}$$

Fick's law, in the form that we require is

$$\frac{dn}{dt} = D\frac{dn}{dx}$$

so that we can write

$$D = \frac{1}{2}q(1 - \varphi)a^2$$

where D is the diffusion coefficient of the charge carriers. Substituting for q gives

$$D = \frac{1}{2}\nu(1 - \varphi)a^2\exp\left[-\frac{E}{kT}\right]$$

We are now able to go back and substitute this into the Einstein relation, to yield

$$\sigma = \frac{\left(ne^2\nu(1 - \varphi)a^2\exp\left[-\frac{E}{kT}\right]\right)}{2kT}$$

This expression is rather cumbersome and it is worth our while to condense it somewhat. To do this let us look at the basic building blocks of the structure that the electrons are diffusing through, that is, the unit cells. If we have c sites that the mobile charge carriers can occupy per unit cell, of volume v, then the number of mobile charge carriers per unit cell will be $c\varphi$, and the number per unit volume will be

$$n = \frac{c\varphi m}{v}$$

where m is the number of unit cells per unit volume. If we now take the vibration frequency to be independent of temperature, we can collect many of these terms into a constant factor K, and write

$$\sigma = K(1 - \varphi)\exp\left[-\frac{E}{kT}\right]$$

where

$$K = \frac{cma^2\nu e^2}{2vk}$$

Appendix 6.2
The Seebeck coefficient and entropy

In order to understand the relationship between the Seebeck coefficient and entropy, it is necessary to consider the effect in terms of the thermodynamics of the system. Electrons, holes or other mobile charge carriers can be considered as chemically reactive species in the thermodynamic sense. Thus, we can allot to them a thermodynamic chemical potential and then use all of the well-established formalism of thermodynamics to derive the relationships that are required. The thermodynamic function which is used is called the *electrochemical potential*, $\bar{\mu}$, defined as

$$\bar{\mu} = \mu + z\Phi$$

where μ is the chemical potential of a mobile charge carrier in the absence of an electrical potential, ze is the charge on the mobile species and Φ is the electric potential in the neighbourhood of the carrier. In the present case, we will consider holes to be the charge carriers of relevance so z will be equal to $+1$ and we can use

$$\bar{\mu} = \mu + e\Phi$$

for holes.

Turning now to a material subjected to a temperature gradient, as illustrated in Figure 6.9, when equilibrium is finally achieved the electrochemical potential of the holes at the hot end of the rod must be equal to that of the holes at the cold end of the rod. Hence we can write

$$\bar{\mu}_H = \bar{\mu}_C$$

$$\mu_H + e\Phi_H = \mu_C + e\Phi_C$$

$$e(\Phi_H - \Phi_C) = \mu_C - \mu_H$$

The chemical potential of a substance can usually be equated to the Gibbs free energy and in this case we can write this as g per hole† so that

$$\mu = g = h - Ts$$

where h and s represent the enthalpy and entropy of a mobile hole and T is the absolute temperature. Using this equation to substitute for μ

$$e(\Phi_H - \Phi_C) = (h_C - T_C s_C) - (h_H - T_H s_H)$$

$$e(\Phi_H - \Phi_C) = h_C - h_H + T_H s_H - T_C s_C$$

We now assume that the enthalpy and entropy of the holes at the hot and cold ends of the material can, to a reasonable approximation, be taken as equal over the small temperature ranges that are normal in experiments. Thus, if we write $h_C = h_H = h$, and $s_C = s_H = s$, and these quantities are substituted into the last equation, it is found that

$$e(\Phi_H - \Phi_C) = (T_H - T_C)s$$

---

†The use of lower case letters g, h and s, indicates that the free energy per particle is being discussed.

$$\frac{(\Phi_H - \Phi_C)}{(T_H - T_C)} = \frac{s}{e}$$

However, we have defined α as

$$\alpha = \frac{(\Phi_H - \Phi_C)}{(T_H - T_C)}$$

so that, for holes

$$\alpha = \frac{s}{e}$$

If we repeat this analysis for electrons we need to substitute the charge on the electron as -1, to give

$$\bar{\mu} = \mu - e\Phi$$

Following through the analysis now gives, for electrons

$$\alpha = -\frac{s}{e}$$

The Seebeck coefficient is really a measure of the entropy of the holes or electrons in the material. At face value this is a rather surprising result and it is worthwhile to check the units to confirm that all is well, at least on that front. The units of entropy will be joules per degree ($J\,K^{-1}$) per particle and the units of the electron charge will be coulombs (C) per particle. Thus the units of the Seebeck coefficient will be

$$(J\,K^{-1}) \text{ per particle/C per particle} = J/CK$$

We have seen previously that

$$J/C = V$$

so that the units of the Seebeck coefficient will be volts per degree ($V\,K^{-1}$), as expected.

Appendix 6.3
The Seebeck coefficient for hopping semiconductors

Suppose that there are n mobile charge carriers in the material. The entropy of these charge carriers can be considered to be due to two parts, the arrangement of the charge carriers, which gives rise to the *configurational* entropy, and the displacements of the charge carriers due to thermal energy, which gives rise to the *vibrational* entropy. We can label these components S_C for configurational entropy and S_V for vibrational entropy. For a material with localized holes or electrons it is possible to estimate S_C by using the Boltzmann formula. This procedure is identical to the calculation of the configurational entropy of point defects in a crystal as set out in chapter 1. Thus we can write

$$S_C = k\ln\left[\frac{c!}{(c-n)!n!}\right]$$

where S_C is the configurational entropy of n particles arranged on c available sites and k is Boltzmann's constant. Using Stirling's approximation (see chapter 1)

$$\ln n! = n\ln n - n$$

we can write

$$S_C = k[c\ln c - n\ln n - (c-n)\ln(c-n)]$$

The entropy per particle, s_C, is given by dS_C/dn, which is

$$s_C = k\ln\left[\frac{(c-n)}{n}\right]$$

However, as we have defined n/c as φ, it is possible to write

$$s_C = k\ln\left[\frac{(1-\varphi)}{\varphi}\right]$$

so that

$$\alpha = -\frac{k}{e}\left\{\ln\left[\frac{(1-\varphi)}{\varphi}\right] + S_V\right\}$$

for electrons and

$$\alpha = +\frac{k}{e}\left\{\ln\left[\frac{(1-\varphi)}{\varphi}\right] + S_V\right\}$$

for holes. In general, S_V is much smaller than $\ln[(1-\varphi)/\varphi]$ and as the only temperature variation will come into this equation from the S_V term, α will be approximately independent of temperature. To a good approximation these equations are of the form

$$\alpha = \pm\frac{k}{e}\left[\ln\left(\frac{n_o}{n_d}\right) + A\right]$$

where A is a constant, n_o is the number of cation sites and n_d is the number of cation defects for materials with non-stoichiometry affecting the metal lattice.

7 Defects and optical properties

7.1 Colour

Non-stoichiometry can have a profound effect on the optical properties of solids, especially colour. This can come about because of the presence of 'coloured' impurities as when the presence of cobalt ions turns normally clear glass blue. Surprisingly, the presence of electronic defects, additional electrons or holes, can also colour materials. This effect is responsible for the colour of the semiprecious gemstones amethyst and smoky quartz.

In both of these cases, and with the others described in this chapter, colour is imparted to a material when electrons interact with *electromagnetic radiation*. When this happens electrons pick up the appropriate amount of energy from the light and are excited from the lower energy *ground state*, E_0, to a higher energy *excited state* E_1, as shown in Figure 7.1(a). The light that is 'left over' is depleted in some frequencies. The reverse can also happen. When electrons drop from the excited state E_1 to the ground state E_0 they release this energy and the same light frequencies will be emitted. This is shown in Figure 7.1(b). The relationship between the energy gained or lost, ΔE, and the frequency, ν, or the wavelength, λ, of the light absorbed or emitted is

$$E_1 - E_0 = \Delta E = h\nu = \frac{hc}{\lambda}$$

where h is Planck's constant and c is the speed of light. For the radiation to be seen, λ is restricted to the visible region of the electromagnetic spectrum, 400–700 nm.

There is an important difference between the interaction of electromagnetic radiation with isolated atoms and with atoms in solids. In isolated atoms, the electrons move between sharp energy levels, E_0 and E_1, as shown in Figure 7.1(a) and (b). This means that light given out corresponds to a sharply defined energy difference and will consist of only one frequency (or a very small range of frequencies). Similarly, only sharp frequencies will be absorbed by gaseous atoms. The absorption or emission spectrum of an atom will consist of a series of sharp lines, each corresponding to one transition, as shown in Figure 7.1(c). In solids, although each electron capable of interacting with light moves between sharp energy levels, lattice vibrations mean that the spacings between the energy levels will vary slightly

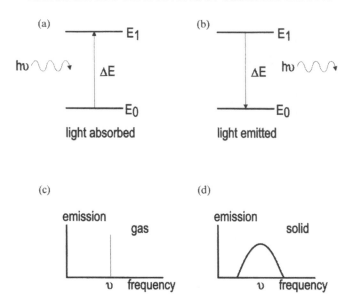

Figure 7.1 The physical processes taking place in light absorption (a) and emission (b). The emission from an isolated gas atom is a sharp line (c) while that from a solid is rather broad (d).

from atom to atom in the crystal. The upshot of this is that there exists a spread of energy levels in the crystal, and the absorption or emission spectrum of a solid due to electron transitions will consist of a series of bell-shaped curves rather than sharp lines, as shown in Figure 7.1(d).

After an electron has been excited into the upper energy level, E_1, as shown in Figure 7.2(a) there are two ways by which the ground state, E_0, can be regained. Normally a light photon will be emitted at random in a process called *spontaneous emission*. When large numbers of photons are emitted from a solid by this process, the light waves are out of step with each other and the light is said to be *incoherent*. This process is shown in Figure 7.2(b). However, there is another way in which light can be given out. If a photon with the exact energy ΔE interacts with the atom while it is in the excited state it can trigger the transition to the ground state, as shown in Figure 7.2(c). This is called *stimulated emission*. Under these circumstances the two light waves are perfectly in step and the light is said to be *coherent*.

Lasers are devices for producing coherent light. Although lasers are more often thought of in terms of high energy output (laser is an acronym for light amplification by stimulated emission of radiation), the coherent nature of the light produced is just as important. In order to make a laser it is necessary to arrange for more atoms to be in the excited state E_2, than there

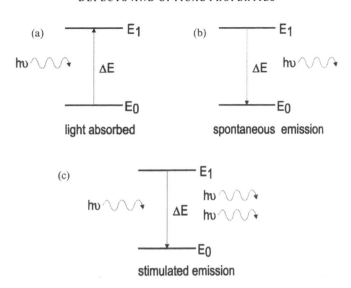

Figure 7.2 Following absorption of radiation (a) an atom can emit radiation by two processes (b) and (c). In (b) the emission takes place at random via spontaneous emission. In (c) the emission is triggered by an incoming photon and leads to stimulated emission. The photons leaving the atom are coherent in this case.

are in the ground state E_1, so that stimulated emission can take place. This state of affairs is called a *population inversion* and it is a necessary precursor to laser action.

In this chapter the way in which defects can be used to make two different sorts of solid state laser will be described.

7.2 Case study: rubies and ruby lasers

Rubies are crystals of Al_2O_3 containing about 0.5% Cr_2O_3 impurity. This composition is a part of the continuous solid solution which is possible between the two end compounds Al_2O_3 and Cr_2O_3. Both compounds share the same *corundum* structure-type shown in Figure 7.3, in which the cations are surrounded by six nearest-neighbour oxygen ions in an octahedral geometry. In the solid solution, the Al^{3+} and Cr^{3+} cations randomly occupy these sites. The lengths of the unit cell axes vary across the solid solution range, in a way quite close to that suggested by Vegard's law, from $a = 0.4763$ nm and $c = 1.3003$ nm for Al_2O_3, to $a = 0.4960$ nm and $c = 1.3599$ nm for Cr_2O_3.

Figure 7.3   The corundum structure. Small filled circles represent Al^{3+} ions and open circles represent O^{2-} ions.

The corundum (Al_2O_3) structure

The corundum structure is adopted by Al_2O_3 (the mineral corundum) and a number of similar oxides such as Cr_2O_3 and Fe_2O_3. The unit cell is hexagonal, with $a = 0.4763$ nm, $c = 1.3003$ nm. There are six Al_2O_3 units in the cell. Each of the cations is surrounded by six oxygen ions in octahedral co-ordination.

It is a surprise to find that the solid solutions $Al_xCr_{2-x}O_3$, with x taking rather small values close to 0.01, are coloured a rich 'ruby' red when the end members are colourless (Al_2O_3) or green (Cr_2O_3). Across the whole solid solution range, the colour is produced by the Cr^{3+} ions. The energy levels involved arise via the interaction of the crystal structure and the d-electrons on the Cr^{3+} ion. This interaction causes the d-electron orbitals to split into two groups, one at a slightly greater energy than the other. This is called the *crystal field* or *ligand field* splitting. An important consequence of this effect is that three new energy levels, E_1, E_2 and E_3 are introduced above the ground state, E_0, which are not present in pure Al_2O_3. These are shown in Figure 7.4(a).

When white light falls onto a crystal of ruby, which is in E_0, Cr^{3+} ions selectively absorb some of the radiation and are excited to energy levels E_2 or level E_3 as shown in Figure 7.4(a). The resulting absorption spectrum, shown in Figure 7.4(b), consists of two overlapping bell-shaped curves, one from E_1 and one from E_3. The absorption curves show that wavelengths corresponding to violet and green–yellow are strongly absorbed. This means that the colour transmitted by the ruby will be red with something of a blue–purple undertone.

What happens to the illuminated ruby crystal which is in an energetically excited state? Some of the Cr^{3+} ions return to the ground state by losing exactly the same amount of energy that was absorbed and so they drop back to the ground state from either E_2 or E_3. Quite a lot of ions, however, lose some energy to the crystal lattice, warming it slightly, and drop back into the energy level labelled E_1. For quantum mechanical reasons, it is not possible for an ion to pass directly from the ground state to E_1 by absorbing energy and so E_1 only gets filled by this round-about process. For the same reason, it is rather difficult for the ion to lose energy in the E_1 state and drop down to the ground state, E_0, again. This means that the return is rather a slow process. Note that this is slow in atomic terms. There are still about 10^2 transitions per second! However, the ions do return to the ground state and in so doing emit light of the appropriate colour. This is also red, but of a slightly different wavelength to that transmitted, as can be seen in Figure

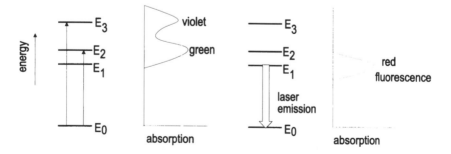

Figure 7.4 (a) Energy is absorbed in a ruby crystal by transitions from the ground state, E_0 to the excited states E_2 and E_3. (b) The spectrum of ruby shows two strong absorption bands near green and violet, due to these transitions. The colour perceived is red with bluish overtones. (c) Laser action occurs when the energy level E_1 is filled and then emptied abruptly by stimulated emission. (d) The laser colour is a red fluorescence and not the normal red colour seen in the gemstone.

7.4(d). The process by which light of a higher energy is absorbed and light of a lower energy is emitted by a solid is called *fluorescence*. The colour of the best rubies is enhanced by this extra red fluorescence.

The ruby solid solution has another surprise in store. At compositions close to $Cr_{0.05}Al_{0.95}O_3$, the crystals are able to support laser action. In fact, ruby crystals were used in the first laser constructed. Laser action comes about in this way. Normally most of the Cr^{3+} ions are in the ground state and only a few are the E_1 energy level. However, if the ruby crystal is illuminated with a very intense light flash, most of the Cr^{3+} ions will absorb energy, become excited into E_2 or E_3 and then pour over into energy level E_1. If the illuminating flash is intense enough to get more Cr^{3+} ions into E_1 than there are in the ground state, E_0, a population inversion has been achieved and the return to the ground state can take place via stimulated emission. The ions remain in the excited state until a photon with an energy ΔE exactly equal to $E_1 - E_0$ enters the ruby. When this happens, it causes an avalanche of all of the Cr^{3+} ions in E_1 to drop to E_0 simultaneously, as shown in Figure 7.4(c). This process makes up laser action, and we have made a ruby laser. The light emitted, a red fluorescence, shown in Figure 7.4(d) is the same as that weakly emitted under normal conditions.

Successful laser action requires that the ions which store the energy, Cr^{3+} in the present case, are well isolated from each other. When the concentration of Cr^{3+} in ruby crystals increases to much more than 0.5%, interaction between these ions allows the energy to be lost in other ways, so preventing laser action from occurring.

So far so good. Al_2O_3 is colourless because there are no Cr^{3+} ions present and we now know that ruby is red because of the extra energy levels introduced along with the Cr^{3+} ions. Can we also explain the green colour of Cr_2O_3? Shouldn't it be red as well? Now it has been found that the magnitude of the ligand field splitting depends very sensitively on the distance between the Cr^{3+} ions and the surrounding octahedron of oxygen ions. In dilute solid solutions, the distance is similar to that in Al_2O_3. As the composition range is spanned, the lattice parameter of the solid solution approaches that of Cr_2O_3 and the position of the critical energy levels E_1 and E_2 alter very slightly. Now the lattice parameter of Cr_2O_3 is larger than that of Al_2O_3. This means that the distance between the Cr^{3+} ions and the surrounding oxygen atoms is larger in Cr_2O_3 than in Al_2O_3 and so the interaction will be smaller. The energy levels E_2 and E_3 will then be a little closer to the ground state than those shown in Figure 7.4(a). This small change is enough to cause the violet and red regions to be strongly absorbed, leaving a green transmission. As the solid solution range is traversed, the colour perceived changes from ruby red through a greyish tone to green.

The colour of rubies depends on the presence of Cr^{3+} substitutional impurities in octahedral sites of a specific geometry. The spinel $Mg(Al_{0.99}-Cr_{0.01})_2O_4$ has 1% Cr^{3+} substituted for Al^{3+} on the octahedral sites. As the geometry of the site is quite similar to that in corundum, the spinel takes on a ruby colour. The colour of this *ruby spinel* is very close to that of real ruby and the two are easily confused visually. In fact, it appears that two of the gemstones in the British Crown Jewels, the *Black Prince's Ruby* and the *Timur Ruby*, are in fact, ruby spinels.

7.3 Transparent electrodes

We will begin this section with a problem and see how the employment of a non-stoichiometric compound has been able to solve it. Many electronic displays, like those that give the answers on calculators or display the time on digital clocks and watches, require electrodes on the front and back of the display. It is necessary for electrodes to have a high electronic conductivity, to prevent excessive power loss and to avoid heating effects. A good electronic conductor needs to have plenty of free electrons and this generally makes for a material that is more or less metallic in appearance, which makes it useless for display electrodes.

Fortunately there are several families of non-stoichiometric oxides which show good electrical conductivity while remaining transparent. The candidates often used are derived from zinc oxide, ZnO, or tin oxide, SnO_2, but here we will only deal with the most widely used material, called indium tin oxide, usually abbreviated to ITO in electronics publications.

Although this material is very widely used, the exact chemical and physical processes taking place which make it into such a good conductor are still not completely understood. However, it is clear that non-stoichiometry has an important role to play. The base material that is used is the insulator indium oxide, In_2O_3. In bulk it is yellow, but in thin films it is quite transparent. To make an electrode a thin film of indium oxide doped with several per cent of tin oxide is laid down on the surface of the device. This is done by heating a mixture of the two oxides in a low partial pressure of oxygen. The oxides evaporate and the molecules condense on the surface of the device which is placed nearby. The film which is deposited is amorphous and the next step is to crystallize it. This is done by heating under carefully controlled conditions of temperature and oxygen partial pressure. At this stage, the tin oxide is incorporated into the indium oxide crystal structure. It is found that the tin occupies indium sites so that impurity substitutional defects are formed.

The consequences of this will depend on which oxide of tin, the dioxide, SnO_2, or the monoxide, SnO, is present. If we have SnO_2 present as the impurity, we will need to incorporate extra oxygen into the In_2O_3. If we have SnO present, we will need to introduce compensating oxygen vacancies into the In_2O_3. In fact, the processing conditions are chosen to have SnO present and so the crystals contain oxygen vacancies. As the film is cooled down, the

Sn^{2+} ions become very unstable and readily give up two electrons each to form Sn^{4+} ions. However, the oxygen diffusion coefficient at room temperature is low and the oxygen vacancies are not filled up as they should be. The released electrons are not trapped at any specific site, but enter the conduction band of the In_2O_3. However, the band gap of the In_2O_3 is not significantly changed by all this and the thin film remains transparent. If the crystals are reheated, oxygen can diffuse in to fill the vacancies. The mobile electrons are used up in this step to form oxygen ions in the crystal matrix. At this stage the material is transparent but reverts to being an insulator again.

7.4   Electrochromic films

Electrochromic materials change colour when subjected to an electric field. In this section we will look at how to use defects in tungsten trioxide to make electrochromic displays. Tungsten trioxide, WO_3, is yellow, an insulator and, like In_2O_3, in thin films it is transparent. Structurally, WO_3 is built of corner-linked WO_6 octahedra and is rather open (see chapter 4). It is quite easy to introduce metal atoms into the cages between the WO_6 octahedra to make non-stoichiometric *tungsten bronzes*. These are most often dark blue–black in colour. The principle of an electrochromic device using tungsten trioxide films is, therefore, not too difficult to envisage. It is necessary to arrange to drive some appropriate metal into the structure using an applied voltage. This will make the tungsten trioxide turn into a blue–black tungsten bronze. Reversal of the voltage must remove the interpolated metal and regenerate the colourless state. This reverse process is often referred to as *bleaching*.

To make such a display we need a reservoir for the interpolated metal and a transparent electrode on the top surface of the display. An experimental display has been constructed using the ionic conductor β-alumina as a source of Na and indium tin oxide films as transparent electrodes. The scheme of the device is shown in Figure 7.5. When the power supply is connected as shown, Na^+ ions migrate into the WO_3 from the β-alumina and electrons enter from the cathode. A tungsten bronze forms which is dark in colour. When the polarity is reversed, the Na re-enters the β-alumina reservoir and the WO_3 becomes colourless once again.

$$WO_3 \text{ (clear)} \underset{-x\text{Na}}{\overset{+x\text{Na}}{\rightleftharpoons}} Na_xWO_3 \text{ (blue–black)}$$

The whole structure is built from non-stoichiometric compounds! The speed of this device depends on ionic diffusion and at ordinary temperatures this is too slow for fast displays such as TV, but is perfect for electronic notice boards or shop signs.

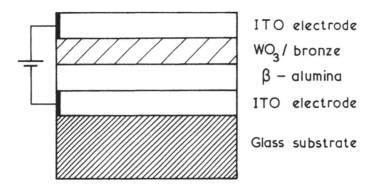

ITO electrode

WO$_3$/ bronze

β – alumina

ITO electrode

Glass substrate

Figure 7.5 An electrochromic display using β-alumina as an Na reservoir, WO$_3$ as the electrochromic film and indium tin oxide (ITO) as transparent electrodes. The display would be viewed from above.

A similar arrangement has been used to make car mirrors which can be electrically dimmed so as to cut down dazzling reflections from bright lights. However, the mirrors have cleverly replaced the β-alumina metal reservoir with something easily available, water vapour! This is decomposed by the voltage supplied to generate H^+ ions which in turn are used to produce a hydrogen tungsten bronze H_xWO_3. The arrangement is shown in Figure 7.6.

On the outer indium tin oxide electrode, the water vapour in the atmosphere is decomposed to hydrogen ions, thus

$$2H_2O \longrightarrow O_2(g) + 4H^+ + 4e^-$$

This is an *electrochemical* decomposition which requires about 1 V at the electrode surface. To drive the protons into the WO$_3$ film we need a proton-conducting electrolyte. One material which is quite a good hydrogen ion conductor is hydrogen uranyl phosphate, $HUO_2PO_4 \cdot 4H_2O$, often referred to as HUP. This material is essentially an acid hydrate and conductivity comes from the easy transport of H_3O^+ via the water molecules in the structure. The next thin film in the cell is, therefore, HUP which is also transparent. The H^+ so produced can pass through the proton-conducting electrolyte to form the bronze, using electrons from the other electrode via the reaction

$$WO_3 \text{ (clear)} \underset{-xH}{\overset{+xH}{\rightleftharpoons}} H_xWO_3(\text{blue–black})$$

The four thin films are laid down on the mirror surface and connected to a battery. If the reflections are too bright, the battery is switched on and the WO$_3$ layer is darkened by the formation of H_xWO_3, and in so doing cuts down the dazzling reflection. When the problem no longer occurs, the

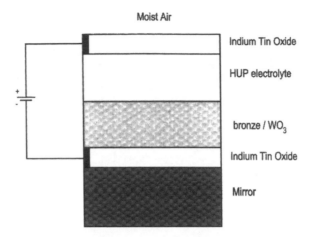

Figure 7.6 An electrochromic layer which can be made to darken on applying a voltage. At the upper indium tin oxide electrode water is decomposed to H^+ which is conducted through the HUP electrolyte to form a hydrogen bronze in the WO_3 layer. Electrons to maintain charge neutrality enter the bronze from the lower indium tin oxide (ITO) electrode. The films are deposited onto the surface of the mirror.

voltage is reversed. The H^+ ions are pulled out from the bronze and the film becomes colourless once more. If a photocell is incorporated into the circuit the whole device can be automated.

Like the sodium tungsten bronze displays, these mirror systems are still in the development stage.

7.5   Case study: the structure of the F-centre

Research in Germany, reported in 1938, indicated that exposure of alkali halide crystals to X-rays caused them to become brightly coloured. In these early studies, the source of the colours were not known and the colour was attributed to the formation of defects that were called *Farbzentrum*, the German for *colour centre*. These defects are now referred to by the briefer title of *F-centres*. Measurement of the absorption spectra of these crystals revealed a more or less bell-shaped curve of the type shown in Figure 7.7. Data for colour centres in alkali halide materials at room temperature are collected in Table 7.1.

Since then it has been found that many different types of high energy radiation, including ultraviolet light, X-rays, γ-rays and neutrons, will cause these F-centres to form. The efficiency of the radiation with respect to F-centre production varies greatly and X-rays, for example, tend to produce F-centres only in the surface layers of the crystal, while the more penetrating γ-

Table 7.1 Alkali metal halide F-centres

Metal	Fluoride		Chloride		Bromide	
	λ_{max} (nm)	Colour	λ_{max} (nm)	Colour	λ_{max} (nm)	Colour
Li	234	colourless	388	colourless	459	blue
Na	344	colourless	459	blue	539	green
K	459	blue	564	green	620	orange
Rb	–	–	620	orange	689	red

rays give a uniform distribution of F-centres throughout the bulk of the material. One significant fact is that, regardless of the type of radiation used, the colour produced in any particular crystal is always the same. Thus, F-centres in NaCl are always an orange–brown colour and in KCl a violet colour, regardless of the method of F-centre production.

An understanding of the true nature of F-centres has involved the correlation of a number of experimental results and the use of a variety of techniques. The first indirect experimental observation that was of interest was that at the same time that F-centres were produced in a crystal, its density fell. This shows that we must, if nothing else, be introducing cation

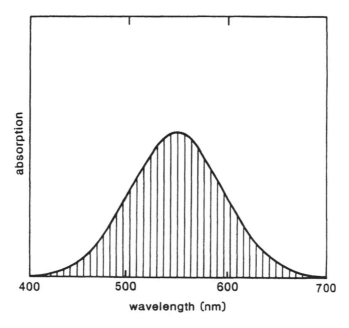

Figure 7.7 A typical bell-shaped absorption curve due to F-centres in KCl. Curves for other F-centres are similar in shape but displaced to other wavelengths.

or anion vacancies into the structure. Now this poses an interesting problem which led to an insight into point defect chemistry which is not, strictly speaking, to do with colour centres themselves but is well worth mentioning.

The fact is that these vacancies cannot be created directly within the body of the crystal because the radiation used is usually not energetic enough to penetrate very far. The problem, therefore, is how to account for the diffusion of the vacancies into the crystal. The diffusion coefficients of both anion and cation vacancies is very low at room temperature, but the colour centres spread considerable distances into the crystals; a movement easily measured with an optical microscope. The difficulty was eventually explained by the suggestion that anion and cation vacancy pairs are the diffusing entities. This is reasonable to us as we have already noted that anion and cation vacancies carry opposite effective charges and hence are likely to associate in pairs. However, at the time this concept was put forward it was a novel idea, and formed the first suggestion that such vacancy pairs could exist. Calculations showed that the enthalpy of migration for a vacancy pair in alkali halide crystals is about $30 \, kJ \, mol^{-1}$ as compared to about $90 \, kJ \, mol^{-1}$ for a cation vacancy and $200 \, kJ \, mol^{-1}$ for an anion vacancy, supporting the contention that pairs of defects are involved.

Despite this success, which seemed able to account for the penetration of colour centres into a crystal, there was still no explanation for the origin of the colour. This is because we know that Schottky defects exist in alkali halides and hence that vacancy pairs will also exist in these materials, but alkali halide crystals are not coloured under normal circumstances. So vacancy pairs themselves cannot be the colour centres.

There are, it turns out, other ways in which we can produce F-centres in alkali halide crystals apart from using ionizing radiation. The first of these involves heating the crystals at high temperatures in the vapour of the alkali metal itself. In a similar way, if we grow crystals of an alkali metal in an atmosphere that contains an excess of alkali metal, colour centres again occur. It is also notable that the exact metal does not matter so long as it is an alkali metal. That is, if we heat a crystal of KCl in an atmosphere of Na vapour the typical purple KCl F-centres are formed, and not the orange NaCl colour centres. Another way of introducing F-centres into alkali metal crystals is to pass an electric current through heated samples and electrolyse them. In this case, the typical F-centre colour is seen to move into the crystal from the cathode region. Once again, the colour depends on the crystal being electrolysed and not the exact nature of the cathode.

These observations suggest that the centres are associated with defects in the crystal structure rather than the exact elements which constitute the compound. Moreover, experiments such as heating crystals in a metal vapour are reminiscent of some of the methods used to produce of non-stoichiometric phases.

What would we expect if this were happening? Consider KCl, for example. If we heat a crystal of KCl in an alkali metal vapour and a little is incorporated into the crystal structure it will occupy normal cation sites as we know that Frenkel defects are not favoured in this compound. We can write the reaction

$$M(g) \longrightarrow M'_K + V^{\bullet}_{Cl}$$

where M stands for the alkali metal added to the KCl crystal. Notice that the M_K carries an effective negative charge because we have added a neutral metal atom to the system and that the vacancy carries the usual effective positive charge.

This state of affairs is not very likely, as alkali metals are very reactive and are more probably found as ions in the crystal. This is easily achieved by liberating the electron from the metal into the crystal. In the transition metal compounds that we discussed earlier this proved to be no problem as the electron could sit at another cation site. Here, though, there is only one plausible site for the electron, and that is at the vacancy. The reaction suggested is:

$$M'_K + V^{\bullet}_{Cl} \longrightarrow M_K + V_{Cl}$$

where the M is now in the normal $1+$ state and the vacancy has an electron trapped at itself, that is

$$V^{\bullet}_{Cl} + e' \longrightarrow V_{Cl}$$

Could this vacancy plus trapped electron be our F-centre? The trapped electron will undoubtedly absorb electromagnetic radiation, and this would lead to an absorption spectrum of the type shown in Figure 7.10. We can check to see if this idea is reasonable by calculating the energy needed to free the electron. We can estimate this energy quite accurately by using the Bohr theory of the hydrogen atom, which gives good numerical answers for the energy of an electron trapped at a positive charge. The theory shows that the energy required to remove an electron completely from the nucleus is given in eV (where 1 eV corresponds to 1.60210×10^{-19} J) by

$$E = -\frac{13.6}{n^2}$$

where the negative sign arises because the energy of the electron is taken as zero when it is free, and becomes increasingly more negative as it approaches the nucleus.

To account for the effect of the crystal lattice, the attractive force must be reduced in proportion to the magnitude of the relative permittivity. Taking this into account the energy becomes

$$E = -\frac{13.6}{\varepsilon_r n^2}$$

where ε_r represents the relative permittivity of the crystal. The energy required is obtained by making n equal to 1. As the magnitude of the relative permittivity for the alkali halides is about 5, we find that the energy to cause the electron to escape from the vacancy is about 2.7 eV. This corresponds exactly to the energy of the F-centre absorption band in NaCl.

There is one more piece of evidence that we can call on that was not available to the earliest investigators. This is provided by the technique of electron spin resonance (esr), which gives a measure of the number of unpaired electrons present in a solid. When this technique is applied to normal alkali metal crystals no unpaired electrons are found, of course. However, for crystals containing F-centres unpaired electrons are found and in numbers equivalent to the number of colour centres present estimated by density measurements. This suggests that each F-centre contains one unpaired electron, as our model has proposed.

This is the end of the trail! An F-centre does indeed consist of a vacancy plus a trapped electron, as is shown in Figure 7.8.

7.6   Electron and hole centres

Since the original studies of F-centres, the term colour centre has broadened in meaning to include any point defect or point defect cluster which have trapped electrons or holes. These are called *electron excess* or *hole excess* centres, respectively.

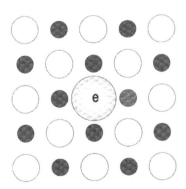

Figure 7.8   The F-centre in an alkali halide crystal. The cations are shown as filled circles and the anions as large open circles. The colour centre consists of an electron trapped at a halide ion vacancy.

The F-centre is an *electron excess centre* and arises because the crystal is slightly non-stoichiometric and contains a small excess of metal. Similar metal excess F-centres exist in compounds other than the alkali halides. Take, for example, an F-centre formed in alkaline earth oxides such as CaO. For this defect to be neutral, two electrons must be trapped as an anion vacancy in an oxide will have an effective positive charge of two units. If only one electron is trapped at such a vacancy it will still retain one unit of effective charge. This centre is called an F'-centre.

A similar defect is found in the mineral *Blue John*.† This is a rare, naturally occurring form of fluorite, CaF_2. The coloration is caused by electron excess F-centres identical to those just described. It is believed that the colour centres were formed by energetic radiation from uranium compounds which were also contained in the rock strata. A fluorite F-centre is shown in Figure 7.9.

One of the best understood hole excess centres gives rise to the colour in smoky quartz and amethyst. These minerals are essentially crystals of silica, SiO_2, which contains a little Al as an impurity. As the aluminium is a trivalent ion which substitutes for silicon in the structure, we need a method of preserving charge neutrality. In natural mineral crystals this is usually by way of incorporated hydrogen, which is present as H^+ in exactly the same amount as the Al^{3+}. The colour centre, giving rise to the smoky purple colour, is formed when an electron is liberated from an $[AlO_4]^{-5}$ group by ionizing radiation and is trapped on one of the H^+ ions present. The reaction can be written as

$$[AlO_4]^{-5} + H^+ \rightarrow [AlO_4]^{-4} + H$$

It is seen that the $[AlO_4]^{-4}$ group is now electron deficient, but once again it is easier to think of this as $[AlO_4]^{-5}$ together with a trapped hole, as we have already done in earlier chapters. The colour arises when the trapped hole absorbs radiation in precisely the same way as the trapped electrons discussed above.

7.7 Case study: the search for a colour centre based information storage medium

7.7.1 Information storage

The ever increasing use of electronic computers has generated a need for sophisticated means of information storage. A reasonably sized book is

---

†The name 'Blue John' is a corruption of the French term 'bleu-jeune'which was used to describe the blue form of the normally yellowish fluorite crystals found in nature.

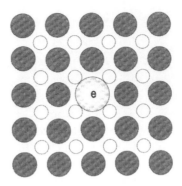

Figure 7.9   The colour centre in fluorite, CaF$_2$, which gives the deep blue colour in the mineral 'Blue John'. It consists of an electron trapped at an F$^-$ vacancy. Ca^{2+} ions are drawn as open circles and F$^-$ ions as large shaded circles.

equivalent to about 10^6–10^7 bits, a figure estimated by assuming that the volume contains, let us say, 50 000 words of six letters each and each letter could be replaced by five bits, using the binary number system with A = 00001, B = 00010, C = 00011 and so on. Surprisingly, a single picture needs the same number of bits to define it. We can confirm this by supposing that the picture can be divided up into about 500 × 500 pixels. Each pixel needs five bits to specify the x-coordinate, another five for the y-coordinate and yet another five to specify the tone or greyness level of the pixel. The total number of bits needed is obtained by multiplying all of these together to give 3.125×10^7 bits! In principle, therefore, a single photographic negative can contain as much information as a book. This comparison illustrates the power of information storage via a *memory plane* and is the reason why the CD-ROM has become the chosen way to store information.

The major advantage of the photographic film is that the high density of information is achieved by virtue of the fact that the x–y location, the geographical position, of each pixel is also stored information, as well as the degree of darkening of each pixel. In addition, each pixel can be read or 'addressed' very rapidly indeed by way of scanned electron or light beams. The major drawbacks of photographic film as an information storage medium are that films need chemical processing to record the information and after this step the information is trapped in an irreversible fashion.

There is, therefore, considerable commercial interest in developing a film in which these drawbacks are eliminated. At the same time, such a film could be used for display screens and similar devices, as well as for information storage. In the following section we describe some research undertaken to try to use colour centres in CaF$_2$ for information storage. Although the work did not result in the development of a commercially successful product, the studies are interesting in their own right and provide an insight into the technical difficulties underlying this type of project.

7.7.2 Photochromic calcium fluoride

The objective is to make a material which will markedly change colour when it is exposed to light. Such materials are called *photochromic*. Ideally, the change should be from white to black. The degree of darkening should also, if possible, be directly proportional to the amount of light incident on the crystal.

The photochromic behaviour of inorganic crystalline materials usually results from the reversible transfer of optically excited electrons from one type of trapping centre, let us call this A, to another type of trapping centre, B as shown in Figure 7.10. When the electron is in site A, then the crystal will be clear, whereas when it is at site B the crystal will be dark. In order to make the changes controllable, it is necessary to use light of one wavelength to transfer electrons from A to B and another wavelength to transfer them back again. This is illustrated schematically in Figure 7.10.

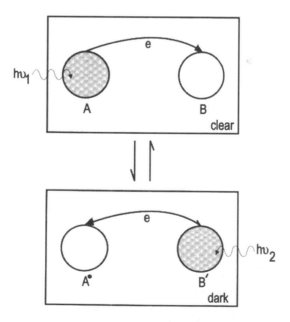

Figure 7.10 Schematic illustration of photochromic behaviour. In (a) the crystal is clear and contains two unionized colour centres, A and B. Transfer of an electron from A to B using radiation of frequency ν_1 gives ionized defects, A^* and B', shown in (b). The crystal in this state is dark, but can be returned to the initial state by radiation of frequency ν_2.

In section 7.5, the F-centres which form in CaF_2 to impart a dark blue–purple coloration were described. The colour is due to the absorption of light which liberates an electron from the F-centre, shown in Figure 7.10. In a normal crystal of CaF_2, the liberated electron will soon be trapped again. The colour centre will remain 'ionized', that is, without its associated electron, only if some other defects can be provided which also act as electron traps. The previous chapter suggests that a multivalent cation would provide a convenient alternative location. All that we need to be sure of is that the cation chosen will form a solid solution with the CaF_2 crystal, and that the trapping of the electron is not so strong as to prevent its return to the F-centre when required. A consideration of the crystal chemistry of CaF_2 shows that a rare-earth cation such as lanthanum, La, would be a suitable second trap.

The initial stage in the preparation of photochromic CaF_2 involves doping the crystals with LaF_3 to create a non-stoichiometric phase. The large La^{3+} ions substitute for Ca^{2+} and occupy normal cation sites in the impure crystal. These form one of the sites needed, the B sites in Figure 7.10.

Reference to chapter 4 will suggest a number of ways in which the crystal can regain charge neutrality. For photochromic purposes, the crystals are especially grown in an atmosphere of HF and He. In these conditions the charge compensation is by way of F^- interstitials which occupy a site next to a substituted La^{3+} cation to form a *defect pair*. The reaction is

$$LaF_3 \xrightarrow{CaF_2} La^{\bullet}_{Ca} + 2F_F + F'_i$$

The colourless crystals are made photochromic by the technique of heating in Ca metal vapour. This process is called *additive coloration*. The Ca metal atoms join the crystal surface and some F^- ions diffuse to the surface to form new anion sites which increases the crystal volume. Now the site rule established in chapter 4 means that for every Ca which joins the crystal two F sites must be created at the surface. The interstitials present will only be sufficient to fill half of these because only one interstitial per La was introduced. Therefore, the net result of the additive doping is to remove all the interstitial F^- ions and in addition create an equal number of F^- vacancies. The halogen vacancies do not occupy random positions but sit in sites next to a La^{3+} ion. This site traps the two electrons provided by the Ca, one to neutralize the effective charge of $+1$ on the neighbouring La and one to neutralize the effective charge of $+1$ on the vacancy. In so doing it forms an F-centre. Because the new centre consists of a halogen vacancy bounded by seven Ca^{2+} ions and one La^{3+} ion it is called an F_{La}-centre as shown in Figure 7.11. The reactions that take place are

$$Ca(g) \xrightarrow{CaF_2} Ca^{2'}_{Ca} + 2V^{\bullet}_F$$

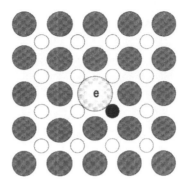

Figure 7.11 An F_{La}-centre in CaF_2. The centre is made up of an La^{3+} ion, shown as a filled circle, located next to an F-centre. Ca^{2+} ions are drawn as open circles and F^- ions as large shaded circles.

$$Ca_{Ca}^{2\prime} + 2V_F^{\bullet} \xrightarrow{CaF_2} Ca_{Ca} + 2V_F$$

$$La^{\bullet} + F^{\prime} + 2V_F \xrightarrow{CaF_2} F_F + (F_{La})$$

It is the F_{La}-centres that form the other sites needed, the A sites in Figure 7.10.

7.7.3 Information storage on defects

Information storage is achieved by swapping electrons between these two defects. In Figure 7.12 the absorption spectrum of the crystals in the initial state is shown shaded. There is little absorption over the visible region, but a strong peak in the near ultraviolet, at about 390 nm, and another in the infra-red, at about 750 nm. The crystals will appear clear. If the crystal is illuminated with light of 390 nm wavelength, the A sites will be 'ionized' and electrons transferred to B sites. The absorption spectrum of the crystal in this state is shown as a dotted line in Figure 7.12. There is now considerable absorption across the visible region and the crystal becomes blue–black. This darkening takes about 2 min at room temperature.

In order to erase the dark colour, the electrons need to be transferred in the opposite direction. The absorption spectrum of the dark state contains a peak at about 680 nm. Light of wavelengths near to this peak will achieve this and illuminating the crystal with this wavelength will transfer the electrons in the reverse direction and turn the crystal clear again. We thus have a system that we can darken or lighten at will. All that is now needed is some way of 'reading' the state of the crystal without altering it. This is achieved by using the absorption band which is present in the clear state at about 720 nm. It is found that irradiation at this wavelength does not alter the degree of darkening present and so monitoring the intensity of this band provides a measure of the state of the system.

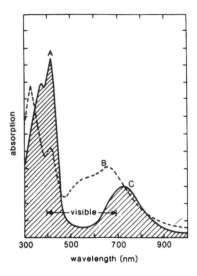

Figure 7.12 The absorption spectrum of CaF_2 doped with LaF_3. The continuous curve represents the clear state and the dotted curve the ionized dark state. Radiation corresponding to peak A is used to turn the crystal dark, i.e. to write information into the crystal. Peak B is used to reverse the process and turn the crystal clear, i.e. to erase information. Peak C is used to read information in the crystal. Radiation of energy corresponding to peak C does not alter the state of the crystal.

To summarize, we have developed an information storage film which does not need chemical processing and is reversible. We can write information into the film using light of wavelengths near to 380 nm, read information stored without changing the information using light of wavelengths near to 720 nm and erase the information using light of about 680 nm. Moreover, the material does not degrade even after many cycles of writing and erasing, and these two processes are very efficient indeed.

This is too good to be true and the immediate question is, if the material is so good, why isn't it being used? Unfortunately, the darkened state is thermally unstable and at normal temperatures lasts for only about a day. In addition, only relatively low concentrations of defects can be introduced into the crystals. This means that to obtain useful changes in optical density rather thick crystals must be used, which increases the amount of material needed to store each bit of information. So the challenge facing scientists and engineers is how to overcome these limitations.

7.8   Colour centre lasers

Although lasers covering a large range of wavelengths are available, not all operate efficiently and there is still a need for others to be developed. For example, in the last section there was a need for lasers operating at three wavelengths to implement the information storage system developed. The fabrication of lasers based on colour centres adds a further dimension to the wavelengths available and several are commercially available. Unfortunately it has been found that ordinary F-centres do not exhibit laser action. In order to achieve laser action it is necessary to create more complex colour centres in which dopant cations are involved. The way in which defects can be manipulated to achieve this objective is described below.

The simplest colour centres that can be used for laser action are F-centres which have one dopant cation next to the anion vacancy. The F_{La}-centres described above fall into this group. Crystals of KCl or RbCl doped with LiCl, containing F_{Li}-centres, shown in Figure 7.13(a), have been found to be good laser materials yielding outputs between 2.45 and 3.45 μm. A unique property of these crystals is that in the excited state an anion adjacent to the F_{Li}-centre moves into an interstitial position, as shown in Figure 7.13(b).

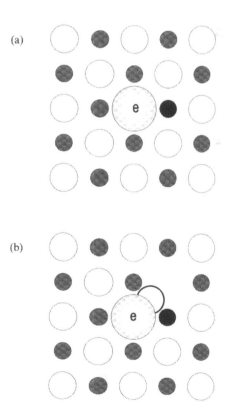

Figure 7.13   (a) A F_{Li}-centre in a cubic alkali halide crystal in the ground state. (b) In the excited state one of the neighbouring anions moves to an interstitial position. This is called type II behaviour and is essential for colour centre laser action.

This is called type II laser behaviour, and the centres are called F_{Li} *(II)* *centres.*

How are these rather complex defects obtained? The process is rather interesting. Take KCl doped with Li as an example. Initially the KCl crystals are grown from a solution containing LiCl as an impurity. The Li^+ cations form substitutional Li_K impurity defects distributed at random throughout the crystal. These defects carry no effective charge as both Li and K are monovalent. The next step is to introduce F-centres into the crystals. This is often carried out by irradiation using X-rays but any of the methods covered in section 7.5 would be satisfactory. The F-centres which form are widely separated and not usually located next to a dopant Li^+ cation.

In order to convert the F-centres into the correct F_{Li} (II)-centres the crystal is subjected to a process called *aggregation*. In this step, the crystals are cooled to about $-10\,°C$ and then exposed to white light. This causes the trapped electrons to be released from the F-centres, leaving normal anion vacancies. These anion vacancies diffuse through the crystal for a while before recombining with the electron once more to reform the F-centre. Ultimately, each F-centre ends up next to a Li^+ ion. At this position it is strongly trapped and further diffusion is not possible. Recombination with an electron forms the F_{Li}-centre required. This process of aggregation is permanent provided that the crystal is kept at $-10\,°C$ and in this state the crystal is laser active.

This rather clever manipulation of defects and defect clusters to produce a new laser is indeed defect engineering!

7.9   Supplementary reading

The best approachable source for an introduction to the topic of colour and colour centres is the book:

K. Nassau, *The Physics and Chemistry of Colour*, chapter 9, Wiley–Interscience, New York (1983).

Bohr's theory of the hydrogen atom is clearly explained by:

H.C. Ohanian, *Physics*, chapter 41, Norton (1985).

Indium tin oxide and other related materials are described in:

S.J. Lynch, *Thin Solid Films* **102**, 47 (1983).

The information on photochromic CaF_2 can be found in the scientific literature. Although these articles are rather advanced reading, they are worth studying to gain an insight into how the material was made and the efforts which were made in order to characterize it. The easiest starting points are:

D.L. Staebler and S.E. Schnatterly, *Phys. Rev. B* **3**, 516 (1971).
W. Phillips and C.R. Duncan, *Metall. Trans.* **2**, 767 (1971).

Colour centre lasers are well described in the manufacturer's literature (see those by Burleigh, for example).
A clear explanation of electrochromic devices is given in:

F.G. Bauke and J.A. Duffy, *Chemistry in Britain* July, 643 (1985).

8 Defects, composition ranges and conductivity

8.1 The equilibrium partial pressure of oxygen over an oxide

The composition of a non-stoichiometric oxide will depend on the partial pressure of oxygen in the surroundings. In this chapter the way in which thermodynamics can help in understanding this will be outlined. Surprisingly, we will find that the number of defects in a material and its electronic conductivity can be quite easily understood from the same viewpoint. We begin by looking at how the stability of an oxide depends on the oxygen pressure.

A good example to take is silver oxide, as this is a stoichiometric phase which decomposes at quite a low temperature, when heated in air, to silver metal and oxygen gas

$$2Ag_2O \longrightarrow 4Ag + O_2$$

The change in the Gibbs free energy for this reaction, ΔG_r, can be related to the partial pressure of the oxygen gas, p_{O_2}, using the equation

$$\Delta G_r = -RT \ln K_p$$

where K_p is the equilibrium constant of the reaction. In this example

$$K_p = p_{O_2}$$

$$\Delta G_r = -RT \ln p_{O_2}$$

At equilibrium, ΔG_r is a constant and so oxygen partial pressure, p_{O_2}, will also be constant. This oxygen pressure is called the *decomposition pressure* or *dissociation pressure* of the oxide, and depends only on the temperature of the system.

What does this equation mean? Suppose some silver metal and silver oxide is sealed in a closed silica ampoule, under a complete vacuum, and we heat the ampoule to a temperature TK. As there is no oxygen in the ampoule some of the silver oxide will decompose and oxygen will be released. This will continue until the equilibrium decomposition pressure is reached. Provided that there is always some silver and silver oxide in the tube the oxygen pressure will be fixed, as shown in Figure 8.1. If the

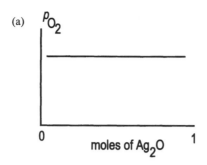

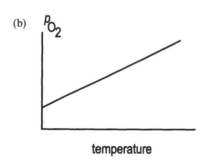

Figure 8.1 (a) Equilibrium between a metal, Ag, and its oxide, Ag$_2$O, generates a fixed partial pressure of oxygen irrespective of the amount of each compound present at a constant temperature. (b) The variation of the partial pressure of oxygen with temperature in a sealed system.

temperature is raised or lowered, either more silver oxide will decompose, or some silver will oxidize, until a new equilibrium decomposition pressure is reached which is appropriate to the new temperature.

The same analysis will hold for any metal–metal oxide mixture. In order to determine the oxygen partial pressure over the pair, it is only necessary to look up the appropriate value of the free energy of formation of the oxide in standard thermodynamic tables at the temperature required and insert the value into the equation

$$\Delta G_r = -RT \ln p_{O_2}$$

This finds use in the laboratory when accurate partial pressures of oxygen must be obtained which are outside the range of conventional pumping systems. A mixture of metal plus oxide is called an *oxygen buffer*. To make an oxygen buffer, a quantity of a metal plus its oxide are placed in the system and heated to the desired temperature. The oxygen partial pressure will reach an equilibrium which depends only on materials chosen.

Example 8.1

Determine the equilibrium oxygen pressure over a mixture of copper, Cu, and cuprous oxide, Cu_2O at 1000 K.
    The chemical equation for the reaction is

$$2Cu_2O \longrightarrow 4Cu + O_2$$

The free energy change for this reaction at 1000 K, ΔG_r, is found to be 188 kJ. Substituting this value in

$$\Delta G_r = -RT \ln p_{O_2}$$

$$\ln p_{O_2} = \frac{-188 \times 10^3}{8.314 \times 10^3}$$

Now these pressures are more often expressed in terms of $\log_{10}$, hence

$$\log p_{O_2} = \frac{-188 \times 10^3}{8.314 \times 10^3 \times 2.303} = -9.8$$

$$p_{O_2} = 1.6 \times 10^{-10} \text{atm}$$

8.2   Variation of partial pressure with composition

A similar analysis to that presented above shows that for any pair of stoichiometric oxides of a multivalent metal M, say M_2O_3 and MO_2, the same relationship must hold. The oxygen pressure in the sealed tube will depend only on the temperature and not on how much of each oxide is present. If we have several oxides in a system, for example MO_2, M_2O_3, M_3O_4 and MO, the partial pressure of oxygen in the system will depend only on temperature and the particular pair of oxides present. However, the pressure will change abruptly when we go from one oxide pair to another, say from M_3O_4–M_2O_3 to M_2O_3–MO_2. This is illustrated schematically in Figure 8.2. Some experimental data for the manganese–oxygen system are presented in Figure 8.3.
    Systems of the sort just discussed, where the oxygen pressure does not depend on the relative amounts of the two phases present but only on the temperature, are referred to as *univariant*. Will the same state of affairs occur with a non-stoichiometric compound? The easiest way to determine this, is to turn to the *phase rule*, which relates the number of components and phases in a system to the number of thermodynamic variables needed to define the state of equilibrium. The 'rule' is usually written as

$$P + F = C + 2 \tag{8.1}$$

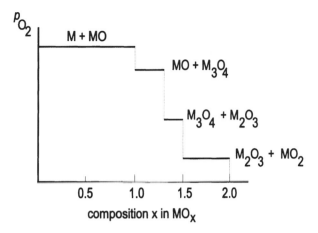

Figure 8.2 The variation of oxygen partial pressure across the metal–oxygen system of a transition metal M which forms stable oxides MO, M_3O_4, M_2O_3 and MO_2.

where P represents the number of *phases* in the system, C is the number of *components* in the system and F represents the minimum number of thermodynamic parameters that have to be specified in order to define the equilibrium completely. This is also called the *variance* of the system.

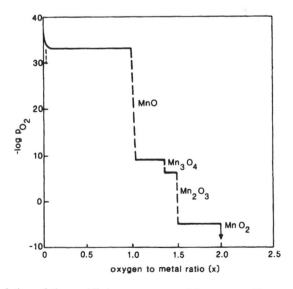

Figure 8.3 Variation of the equilibrium oxygen partial pressure with composition for the oxides of manganese. The full lines parallel to the composition axis represent two-phase regions and the dashed lines parallel to the $\log p_{O_2}$-axis represent single-phase regions.

In the case of silver oxide, equilibrium is achieved when *three phases* coexist, Ag_2O, Ag and O_2. The number of components in the system is two, silver and oxygen. Hence, substituting this information into the equation 8.1, we find

$$3 + F = 2 + 2$$

i.e. $F = 1$.

What does this tell us? It tells us that under conditions where we have three phases in equilibrium we can only vary one thermodynamic parameter, all the others are fixed. We have already shown this to be true. If we choose the temperature of our sealed ampoule, then the oxygen pressure is fixed. Similarly, if we choose pressure, only one temperature will allow the system to come to equilibrium. Thus, the system has a variance of one; it is *univariant*, as stated earlier.

Now consider a non-stoichiometric compound, $\sim MO$, in equilibrium with the vapour phase. If we change the composition of the compound a little, we do not produce more of a second solid compound, but still have one solid in equilibrium with gas. The number of components in the system will be two, metal and oxygen and the number of phases will be two, $\sim MO$ and O_2, gas, hence

$$P + F = C + 2$$

$$2 + F = 2 + 2$$

$$F = 2$$

The minimum number of thermodynamic parameters that have to be fixed to define the equilibrium is now two, and the system is said to be *bivariant*. In this experiment, if $\sim MO$ and O_2 gas are present in a sealed tube at a certain temperature, we cannot be sure that we are at equilibrium. Another parameter must be specified to define the system completely. This could be oxygen pressure or the composition of the oxide. Thus, if we specify oxide $MO_{1.0980}$ is in contact with O_2 gas at TK, the equilibrium oxygen partial pressure will be defined. Similarly, if we define the oxygen partial pressure and the temperature, the composition of the phase will have one precise value only.

A system which appears to contain only one solid phase and which shows bivariant behaviour must contain a non-stoichiometric compound. This serves as a *thermodynamic definition* of non-stoichiometry to complement the structural one given in chapter 4.

Let us illustrate this behaviour for the Fe–O system, heated at a fixed temperature. If we start by oxidizing iron metal in a sealed tube, the first pair of equilibrium products will be Fe coexisting with the lower composition range of $\sim$FeO, FeO$_{1.050}$. In this composition range, the oxygen partial pressure will be constant, as shown in Figure 8.4. Similarly, Fe$_3$O$_4$ will be in equilibrium with the oxygen-rich end of the $\sim$FeO phase, that is, FeO$_{1.150}$. The partial pressure over the oxides will also be constant, as shown in the Figure 8.4. However, within the stoichiometry range of $\sim$FeO, between the composition limits of approximately FeO$_{1.050}$ and FeO$_{1.150}$, only one solid phase will be present. The system will now be bivariant in behaviour and the partial pressure of oxygen over the $\sim$FeO will depend on the composition. This is shown as a sloping line in Figure 8.4. Changing the temperature will change the details on this figure but the overall shape of the diagram will be the same.

8.3   Electronic conductivity and partial pressure for Ni$_{1-x}$O

Cation deficient oxides which accommodate their non-stoichiometry by way of cation vacancies are typified by Ni$_{1-x}$O. In this oxide, the cation vacancies are balanced by compensating Ni^{3+} ions to maintain charge balance, and the formula can be written as Ni$^{2+}_{1-3x}$Ni$^{3+}_{2x}$O. We can write down a chemical equation for the production of vacancies and Ni^{+3} ions following the guidelines set out in chapter 4

$$2Ni_{Ni} + \frac{1}{2}O_2(g) \xrightarrow{NiO} O_O + V''_{Ni} + 2Ni^{\bullet}_{Ni} \tag{8.2}$$

where V''_{Ni} represents a cation vacancy with two virtual negative charges and

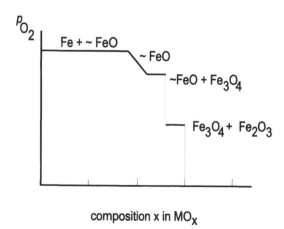

composition x in MO$_X$

Figure 8.4   Schematic illustration of the variation of oxygen partial pressure, plotted as log p_{O_2} versus composition, x, for the iron–oxygen system. The heavy lines represent two oxides present at equilibrium and the diagonal line represents the non-stoichiometric $\sim$FeO phase. The $\sim$FeO phase range has been exaggerated for clarity.

$Ni^{\bullet}_{Ni}$ is an Ni^{2+} cation on a normal Ni^{2+} site, but bearing an extra positive charge, that is, an Ni^{3+} ion. If there is some uncertainty that the charge is located on normal Ni^{2+} ions, we may prefer to write the equation as

$$\frac{1}{2}O_2(g) \xrightarrow{\text{NiO}} O_O + V^{2\prime}_{Ni} + 2h^{\bullet}$$

where $h^{\bullet}$ represents a positive hole, not trapped or located at any particular site in the lattice, but free to move through the valence band.

Having expressed the defect formation reaction in terms of a chemical equation, we can handle it by normal equilibrium thermodynamics. The equilibrium constant of reaction (8.2) is

$$K = \left[\frac{[Ni^{\bullet}_{Ni}]^2 [V^{2\prime}_{Ni}][O_O]}{[Ni_{Ni}]p_{O_2}^{1/2}} \right]$$

where [ ] represent concentrations, and p_{O_2} is the oxygen partial pressure. Now the values of $[O_O]$ and $[Ni_{Ni}]$ are essentially constant, as the change in stoichiometry is small. Hence we can assimilate them into a new constant, K_1 and write

$$K_1 = \left[\frac{[Ni^{\bullet}_{Ni}]^2 [V^{2\prime}_{Ni}]}{p_{O_2}^{1/2}} \right]$$

We also know that for every vacancy in the crystal we have two Ni^{3+} ions, to maintain electroneutrality, so that

$$[V^{2\prime}_{Ni}] = \frac{1}{2}[Ni^{\bullet}_{Ni}]$$

Hence

$$K_1 = \frac{[Ni^{\bullet}_{Ni}][Ni^{\bullet}_{Ni}][Ni^{\bullet}_{Ni}]}{2p_{O_2}^{1/2}}$$

so that the concentration of Ni^{3+} ions is given by

$$[Ni^{\bullet}_{Ni}] = (2K_1)^{1/3} p_{O_2}^{1/6}$$

The concentration of Ni^{3+} ions is, therefore, proportional to the *1/6 power of the oxygen partial pressure*. Alternatively, the concentration of holes, $[h^{\bullet}]$, is proportional to the 1/6 power of the oxygen partial pressure, i.e.

$$h^{\bullet} \propto p_{O_2}^{1/6}$$

The electrical conductivity, σ, will be proportional to the concentration of holes, so we can write

$$\sigma \propto p_{O_2}^{1/6}$$

This is a very interesting result. It tells us that the conductivity will increase with increasing oxygen pressure. Moreover, the pressure dependence is specific. The conductivity will increase as $p_{O_2}^{1/6}$. Some experimental results confirming this result are shown in Figure 8.5.

8.4   Case study: $Co_{1-x}O$

The results obtained for $\sim$NiO are rather remarkable. The pressure dependence of conductivity was precisely that expected from the defect model used. This suggests that the technique could be used to clarify the nature of the defects present in other systems. The oxide $Co_{1-x}O$ provides a good illustration of this. It would be expected that this oxide would behave in a very similar way to $\sim$NiO, its periodic table neighbour. However, experimental results, shown in Figure 8.6, show that the conductivity increases as

$$\sigma \propto p_{O_2}^{1/4}$$

Can a model of the defects present be derived which accounts for this?

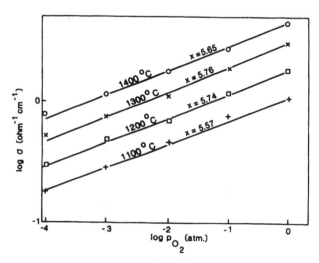

Figure 8.5   Plot of conductivity versus the logarithm of oxygen partial pressure for single, high purity nickel oxide crystals determined at a variety of temperatures. The slopes of the lines, $1/x$, are close to, but not exactly equal to, 1/6. [Data redrawn from C.M. Osborn and R.W. Vest, *J. Phys. Chem. Solids* **32**, 1131, 1343, 1353 (1971).]

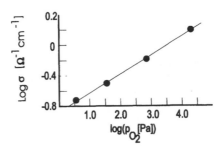

Figure 8.6 Variation of the conductivity, at 950 °C, as a function of the partial pressure of oxygen for ~CoO. [Data redrawn from C. Kowalski, thesis, University Henri Poincaré, Nancy I (1994).]

The crux of the analysis of ~NiO was that the two holes generated by each cation vacancy sat on different Ni^{2+} cations. Could it be that instead of two holes sitting on separate cations to make two Co^{3+} ions, both sit on one cation to form a Co^{4+} ion?

Let us analyse this by following the same steps as in the case of ~NiO. The chemical equilibrium equation is

$$Co_{Co} + \frac{1}{2}O_2(g) \xrightarrow{CoO} O_O + V_{Co}^{2\prime} + Co_{Co}^{2\bullet}$$

or

$$\frac{1}{2}O_2(g) \xrightarrow{CoO} O_O + V_{Co}^{2\prime} + (2h^{\bullet})$$

where the two holes are bracketed together to show that they are associated at one site. Following through the analysis as before, we find the equilibrium constant, K, is given by

$$K = \left[\frac{[O_O][V_{Co}^{2\prime}][Co_{Co}^{2\bullet}]}{[Co_{Co}]p_{O_2}^{1/2}} \right]$$

Taking the concentrations of normal Co^{2+} and O^{2-} ions as constant because CoO has a small stoichiometry range, and incorporating them into a new equilibrium constant K_1

$$K_1 = \left[\frac{[V_{Co}^{2\prime}][Co_{Co}^{2\bullet}]}{p_{O_2}^{1/2}} \right]$$

Also $[V_{Co}^{2\prime}]$ is equal to $[Co_{Co}^{2\bullet}]$ and hence

$$[V_{Co}^{2\prime}] = [Co_{Co}^{2\bullet}] = [2h^{\bullet}] = K_1 p_{O_2}^{1/2}$$

$$[2h^\bullet] = K_1^{1/2}p_{O_2}^{1/4}$$

$$\sigma \propto p_{O_2}^{1/4}$$

This predicts p-type semiconductivity, which increases with oxygen pressure, but now the dependence is proportional to the 1/4 power rather than the 1/6 power. This seems like a very successful outcome. It appears that the defects in $Co_{1-x}O$ are cation vacancies and Co^{4+} ions.

It is possible to take this just a little further. Chemically, it is possible to make the suggestion that two short-lived Co^{3+} ions disproportionate into one Co^{2+} ion and one Co^{4+} ion at the temperature of the experiment

$$2Co^{3+} \xrightarrow{CoO} Co^{2+} + Co^{4+}$$

Perhaps this reaction does not take place so readily at higher temperatures. Experimentally this could be checked by following the conductivity as a function of oxygen pressure at different temperatures. If the pressure dependence tends to change towards $+1/6$ the idea would be strongly supported. This is found. Although the change is rather small, as can be seen from Figure 8.7, there is no doubt about the result. At a temperature of 1150 °C the partial pressure dependence is $+1/4.16$. If 100% Co^{3+} defects give a slope of 1/6 and 100% Co^{4+} defects give a slope of 1/4, $y\%$ of Co^{3+} will give a slope of 1/4.16 where

$$y\left(\frac{1}{6}\right) + (1-y)\left(\frac{1}{4}\right) = \frac{1}{4.16}$$

The solution is $y = 11.5\%$. It thus appears that 11.5% of the Co^{4+} ions

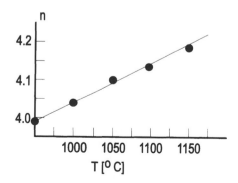

Figure 8.7 The variation of the exponent, n, where conductivity $\sigma \propto p_{O_2}^{1/n}$, as a function of temperature for $\sim$CoO. [Data redrawn from C. Kowalski, thesis, University Henri Poincaré, Nancy I (1994).]

have disproportionated at $1150\,^\circ$C and it is possible to speculate that even more Co^{3+} would form at even higher temperatures. The theory has been rather successful.

8.5   Electronic conductivity and partial pressure for $Zn_{1+x}O$

In order to illustrate the behaviour of oxygen deficient materials let us look at oxides which have interstitial cations in the structure and which are n-type semiconductors, typically ZnO. When ZnO is heated in zinc vapour, we obtain a non-stoichiometric crystal containing excess zinc, $Zn_{1+x}O$. The reaction is

$$Zn(g) \xrightarrow{ZnO} Zn_i^{\bullet} + e'$$

where, for the moment, it has been assumed that the interstitial zinc atoms are singly ionized to form Zn^+ ions. The equilibrium constant for this reaction is

$$K = \frac{[Zn_i^{\bullet}][e']}{p_{Zn}}$$

The concentration of the Zn interstitials is equal to that of the electrons and is given by

$$[Zn_i^{\bullet}] = [e'] = K^{1/2}p_{Zn}^{1/2}$$

Thus the number of defects increases as the vapour pressure of zinc metal increases, and the semiconductivity will be proportional to the zinc vapour pressure following the equation

$$\sigma \propto p_{Zn}^{1/2}$$

We can treat the oxygen dependence similarly. In this case, heating ZnO in a vacuum will cause oxygen to be lost from the crystal, to again produce $Zn_{1+x}O$. Thus, we can write the chemical reaction as

$$O_O \xrightarrow{ZnO} Zn_i^{\bullet} + e' + \frac{1}{2}O_2(g)$$

The equilibrium constant for this reaction is given by

$$K = \frac{[Zn_i^{\bullet}][e']p_{O_2}^{1/2}}{[O_O]}$$

If we follow previous procedures and amalgamate the concentrations of lattice oxygen, which will be large and almost constant, into the equilibrium constant K to generate a new constant, K_1, we can rewrite this equation as

$$K_1 = [Zn_i^{\bullet}][e']p_{O_2}^{1/2}$$

As before

$$[Zn_i^{\bullet}] = [e']$$

so that

$$[Zn_i^{\bullet}] = [e'] = K_1^{1/2}p_{O_2}^{-1/4}$$

This shows that the number of defects will be proportional to the $-1/4$ *power of the oxygen partial pressure*. The number of defects will fall as the oxygen pressure increases. As the oxygen pressure increases the composition of the material will approach the stoichiometric formula of $ZnO_{1.000}$. The electronic conductivity is proportional to the number of electrons so

$$\sigma = p_{O_2}^{-1/4}$$

It is seen that the situation here is opposite to that encountered in the p-type oxides and now the conductivity will decrease as the oxygen partial pressure increases. Some classical experimental data for $Zn_{1+x}O$, shown in Figure 8.8, indicate that the relationship is obeyed well. This in turn supports the original assumption that the interstitial Zn atoms are singly ionized and exist as Zn^+ entities in the structure.

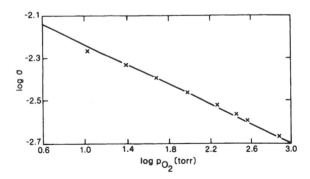

Figure 8.8 Classical data showing the conductivity of $\sim$ZnO as a function of oxygen pressure at 650 °C as determined by Baumback and Wagner in 1933. The data are evidence that the Zn_i atoms are singly ionized rather than doubly ionized.

Example 8.2

What would the Zn pressure dependence of conductivity be if the zinc interstials in ZnO were doubly ionized?

The chemical reaction describing the formation of doubly ionized interstitials is

$$Zn(g) \xrightarrow{ZnO} Zn_i^{2\bullet} + 2e'$$

If the analysis is now followed through in the same way as above the result is

$$\sigma \propto p_{Zn}^{1/3}$$

8.6   Brouwer diagrams

Many of the important semiconducting materials show composition ranges which take them from cation deficient to cation excess. At the same time the semiconductivity often changes from p-type to n-type or vice versa. From a practical point of view, it would be helpful to be able see how the conductivity will change with composition at a glance without always having to return to chemical equilibrium equations. *Brouwer diagrams* allow one to do just that. The best way to understand the information contained in these diagrams is to construct an example. In the rest of this section a Brouwer diagram for a non-stoichiometric compound with a composition close to MX will be constructed. In this example it will be assumed that both M and X are divalent.

8.6.1   Initial assumptions

The first step is to set out the assumptions concerning the defects that are likely to occur, using physical and chemical intuition about the system in mind. (Remember that the diagram is to be useful in the laboratory.) In this example, we shall presume that:

1.   only vacancies are important in $\sim MX$, and interstitial defects can be ignored
2. $\sim MX$ can have an existence range which spans both sides of the stoichiometric composition, $MX_{1.00}$
3.   the electrons or holes in the non-stoichiometric compound are not trapped at the vacancies, but are free to move
4.   the most important gaseous component is X_2, as is the case in most oxides, halides and sulphides.

These assumptions mean that there are only four defects to consider, electrons, e', holes, $h^•$, vacancies on metal sites, $V_M^{2\prime}$ and vacancies on anion sites, $V_X^{2•}$.

8.6.2 Defect equilibria

It is now necessary to set up chemical equations to describe the equilibrium between these defects. These are:

1. the creation and elimination of Schottky defects. These defects can form at the crystal surface or vanish by diffusing to the surface. The equation describing this is

$$\text{zero} \longrightarrow V_M^{2\prime} + V_X^{2•}$$

2. the creation and elimination of electronic defects. Electrons can combine with holes to be eliminated from the crystal thus

$$\text{zero} \longrightarrow e' + h^•$$

3. the composition can change by interaction with the gas phase to produce cation vacancies

$$\frac{1}{2}X_2 \xrightarrow{MX} X_X + V_M^{2\prime} + 2h^•$$

or anion vacancies

$$X_X \xrightarrow{MX} \frac{1}{2}X_2 + V_X^{2•} + 2e'$$

4. the electrical neutrality must be maintained

$$2\left[V_M^{2\prime}\right] + \left[e'\right] \rightarrow 2\left[V_X^{2•}\right] + \left[h^•\right] \tag{8.3}$$

This latter is the *key equation*.

8.6.3 Equilibrium constants

The equilibrium constants of these equations are then written down

$$K_S = \left[V_M^{2\prime}\right]\left[V_X^{2•}\right] \tag{8.4}$$

$$K_e = \left[e'\right]\left[h^•\right] \tag{8.5}$$

$$K_3 = \frac{[V_M^{2'}][h^\bullet]^2}{p_{X_2}^{1/2}} \tag{8.6}$$

$$K_4 = [V_X^{2\bullet}][e']^2 p_{X_2}^{1/2} \tag{8.7}$$

K_4 is actually redundant, as

$$K_4 = \frac{K_S K_e^2}{K_3}$$

but it simplifies things to retain it.

There is now (almost) enough information available to draw the diagram.

8.6.4   High X_2 partial pressures

Under these conditions it is unlikely that there will be a high population of anion vacancies and so an assumption is made that cation vacancies predominate. The discussion earlier in this chapter indicates that cation vacancies are usually paired with positive holes and so the it is assumed that there are more holes than electrons present. The appropriate form of equation (8.3) for the high pressure region is

$$2[V_M^{2'}] = [h^\bullet] \tag{8.8}$$

We can now substitute into equations (8.4)–(8.7) to obtain relationships between the partial pressure of X_2 and the defect concentrations present in the material. Starting with equation (8.6)

$$[V_M^{2'}]\left(2[V_M^{2'}]\right)^2 = K_3 p_{X_2}^{1/2}$$

$$4[V_M^{2'}]^3 = K_3 p_{X_2}^{1/2}$$

$$8[V_M^{2'}]^3 = 2K_3 p_{X_2}^{1/2}$$

so that

$$[V_M^{2'}] = \frac{1}{2}(2K_3)^{1/3} p_{X_2}^{1/6}$$

Substituting from equation (8.8)

$$[h^\bullet] = (2K_3)^{1/3} p_{X_2}^{1/6} \tag{8.9}$$

From equation (8.5)

$$[e'] = \frac{K_e}{[h^\bullet]}$$

i.e.

$$[e'] = \frac{K_e}{(2K_3)^{1/3} p_{X_2}^{1/6}}$$

From equation (8.4)

$$[V_X^{2\bullet}] = \frac{K_S}{[V_M^{2\prime}]}$$

$$[V_X^{2\bullet}] = 2K_S(2K_3)^{-1/3} p_{X_2}^{-1/6}$$

The equilibrium constant K_4 is not needed here.

To display these results graphically take logarithms of each side of the four equations describing the defect populations. For example, with equation (8.9)

$$[h^\bullet] = (2K_3)^{1/3} p_{X_2}^{1/6} \qquad (8.9)$$

$$\log[h^\bullet] = \frac{1}{3}\log(2K_3) + \frac{1}{6}\log p_{X_2}$$

A plot of $\log[h^\bullet]$ against $\log p_{X_2}$ is a straight line of slope 1/6. Linear relationships between $\log$ [defect] and $\log p_{X_2}$ result for all four equations, as shown in Figure 8.9.

Figure 8.9 contains all of the information given in the equations written out above but in an accessible way. The key defect equation

$$2[V_M^{2\prime}] = [h^\bullet]$$

is displayed prominently at the top of the diagram to remind us of the main assumption. Most obviously, it is possible to see that there are only four defect types present as there are only four defect lines shown. A glance at the diagram shows that holes predominate so that the material is a p-type semiconductor. In addition, the conductivity will increase as the 1/6 power of the partial pressure of the gaseous X_2 component increases. The number of metal vacancies (and oxygen excess) will increase as the partial pressure of the gaseous X_2 component increases. This diagram could be used for the case of $Ni_{1-x}O$ described above.

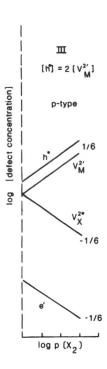

Figure 8.9 Partial Brouwer diagram for a phase $\sim MX$ in the region where the partial pressure of the gaseous component X_2 is high. The logarithm of the defect concentrations, $\log[h^\bullet]$, $\log[V_M^{2\prime}]$, $\log[V_X^{2\bullet}]$ and $\log[e^\prime]$ are plotted along the y-axis, versus the logarithm of the partial pressure of X_2 along the x-axis. The slopes of each of the lines is $\pm 1/6$, and the material will be a p-type semiconductor over the whole region covered.

8.6.5 Medium X_2 partial pressures

As the partial pressure of X_2 decreases, the number of cation vacancies and holes will decrease as the composition of $\sim MX$ approaches $MX_{1.000}$. Stoichiometric crystals tend to be insulators, and hence it is more appropriate to suppose that the formation of vacancies on cation and anion sites is more important than the creation of electrons and holes. It is now more reasonable to approximate the key equation (8.3)

$$2[V_M^{2\prime}] + [e^\prime] \rightarrow 2[V_X^{2\bullet}] + [h^\bullet] \tag{8.3}$$

by the new key relation

$$2[V_M^{2\prime}] = 2[V_X^{2\bullet}]$$

This equation is now substituted into equations (8.4)–(8.7) to obtain a new set of equations for the defect concentrations

$$\left[V_M^{2\prime}\right] = K_S^{1/2}$$

$$\left[V_X^{2\bullet}\right] = K_S^{1/2}$$

$$\left[h^\bullet\right] = \left(\frac{K_3}{K_S^{1/2}}\right)^{1/2} p_{X_2}^{1/4}$$

$$\left[e^\prime\right] = K_e \left(\frac{K_3}{K_S^{1/2}}\right)^{-1/2} p_{X_2}^{-1/4}$$

We can plot these on the log concentration versus log p_{X_2} graph, as before, to produce the result shown in Figure 8.10.

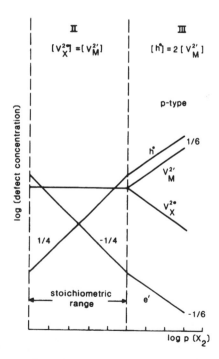

Figure 8.10 Partial Brouwer diagram for a phase $\sim MX$, extended to lower partial pressures than applicable in Figure 8.10. The fraction besides the lines represents the slope of the lines. As the numbers of cation and anion vacancies is equal in the region to the left of the dotted line, the material in this part of the diagram has a composition $MX_{1.0}$.

What does the diagram show? In the new region on Figure 8.10 there are only three defect lines because the cation and anion vacancy equations overlap. The number of holes and electrons will be well below the number of cation and anion vacancies for most of this region. The material will be a stoichiometric insulator with a composition $MX_{1.00}$ containing Schottky defects. At the extremes of the range the material becomes p-type with a pressure dependence of $+1/4$ or n-type with a pressure dependence of $-1/4$.

8.6.6   Low X_2 partial pressures

If we keep decreasing the partial pressure of X_2, anion vacancies would be expected to dominate. The defect equation (8.3)

$$2[V_M^{2\prime}] + [e'] \rightarrow 2[V_X^{2\bullet}] + [h^\bullet]$$

can now be approximated by the new key equation

$$[e'] = 2[V_X^{2\bullet}]$$

This is substituted into equations (8.4)–(8.7) to derive the new defect concentrations.

$$[V_X^{2\bullet}] = \frac{1}{2}\left(\frac{2K_S K_e^2}{K_3}\right)^{1/3} p_{X_2}^{-1/6}$$

$$[e'] = \left(\frac{2K_S K_e^2}{K_3}\right)^{1/3} p_{X_2}^{-1/6}$$

$$[h^\bullet] = \left[\frac{K_2}{(2K_1 K_2^2/K_3)^{1/3}}\right] p_{X_2}^{1/6}$$

$$[V_M^{2\prime}] = \left[\frac{2K_S}{(2K_S K_e^2/K_3)^{1/3}}\right] p_{X_2}^{1/6}$$

These equations can now be plotted as straight lines if we take logarithms of both sides. The result is shown in Figure 8.11, plotted as region I.

8.6.7   The complete diagram

In Figure 8.11, the whole Brouwer diagram for the system is shown. There are three regions corresponding to low, medium and high partial pressures of X_2 gas. The electron concentration starts high in the n-type region I and falls progressively, while the hole concentration starts low and ends high in

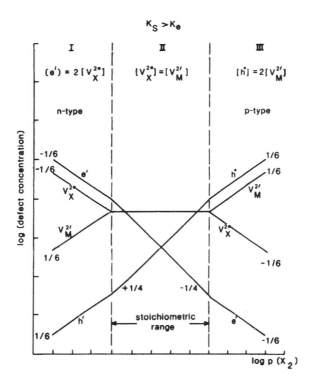

Figure 8.11 Complete Brouwer diagram for a phase ~MX for conditions of equilibrium specified by the electroneutrality equations shown in the three regions I, II and III.

the p-type region III. The way in which the other defects change as the partial pressure increases is easy to see. This variation gives a number of clues about the way in which the properties of the phase change as the partial pressure of X_2 and the composition alter across the diagram. On the left, for example, there is high concentration of anion vacancies and so easy diffusion of anions is to be expected. In region III, we have a high concentration of cation vacancies which would be expected to enhance cation diffusion.

8.6.8 Further considerations

The form of this diagram can easily be modified to take into account other defect types and equilibria. For example, a treatment of AgBr would include Frenkel defect formation and not Schottky defects. It would also be easy to smooth out the abrupt changes between the three regions I, II and III by including an intermediate key electroneutrality equation. Between regions II and III this could be

$$2[V_M^{2\prime}] = 2[V_X^{2\bullet}] + [h^\bullet]$$

Similarly, between regions I and II one could use

$$2[V_M^{2\prime}] + [e\prime] = [V_X^{2\bullet}]$$

These variations will be seen in Brouwer diagrams for many systems of importance, especially those which refer to semiconducting materials.

8.7   Case study: an experimentally determined Brouwer diagram; CdTe

The material cadmium telluride, CdTe, is an important semiconductor. It crystallizes in the *sphalerite* (also called the *zinc blende*) structure. It has a (temperature-dependent) composition range from $CdTe_{1.0000}$ to approximately $CdTe_{1.0204}$. It is an excellent photoconductor and the resistance of the material shows a large decrease when the compound is illuminated. This leads to its use in photodiodes and other light-detecting systems.

In order to prepare device quality crystals, it is important to control the defect population precisely. The Brouwer diagram, shown in Figure 8.13, can help in this. On the vertical axis, the concentrations per cubic centimetre of the most important defects are shown. On the pressure axis ($K_r p_{Cd}$) is shown. This relates to the important reaction

$$Cd(g) \xrightarrow{CdTe} Cd_i^\bullet + e\prime$$

by which Cd atoms in the vapour enter the crystal as single-charged interstitials and liberate an electron to the lattice. The equilibrium constant for this reaction, K_r, is given by

$$K_r = \frac{[e\prime][Cd_i^\bullet]}{p_{Cd}}$$

What information does the figure contain? First of all we can see that there are seven defects considered to be of importance, holes, $h^\bullet$, electrons, $e\prime$, neutral cadmium interstitials, Cd_i, cadmium interstitials which have an effective charge of $+1$, $Cd_i^\bullet$, and three sorts of vacancies on cadmium sites, double charged, $V_{Cd}^{2\prime}$, singly charged, $V_{Cd}\prime$, and neutral vacancies, V_{Cd}.

Consider the central part of the diagram, region II, first. The basic point defect equilibria that holds must be a Frenkel type on the Cd sub-lattice, because of the presence of cadmium interstitials and vacancies. Stoichiometric $CdTe_{1.00}$ is formed when the number of Cd interstitials equals the number of Cd vacancies, that is, when

$$[Cd_i^\bullet] = [V_{Cd}\prime]$$

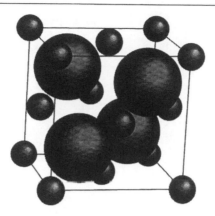

Figure 8.12 The sphalerite (zinc blende) structure. The structure contains alternating layers of Zn and S atoms. Each atom is surrounded tetrahedrally by atoms of the opposite type.

The sphalerite (zinc blende) structure

The sphalerite or zinc blende structure type of ZnS is similar to that of diamond but with alternating layers of Zn and S replacing C, as shown in Figure 8.12. The cubic cell has a lattice parameter of $a = 0.541$ nm. There are four Zn and four S atoms in a unit cell.

which only occurs at a precisely defined pressure of Cd vapour, marked S on the x-axis of the figure. If it is necessary to fabricate stoichiometric CdTe, this pressure must be used. In this material, the Frenkel defect equilibrium is much less important than the electronic defects present, so that region II is defined by the equation

$$[e'] = [h^{\bullet}]$$

A glance at the diagram shows that the number of Frenkel defects present is about 10^{12} cm^{-3} compared to about 10^{16} cm^{-3} for holes and electrons. The holes and electrons are present in equal numbers and the material will be an *intrinsic* semiconductor.

In region I, corresponding to low partial pressures of Cd, the charge neutrality condition was taken as

$$[h^{\bullet}] = [V'_{Cd}]$$

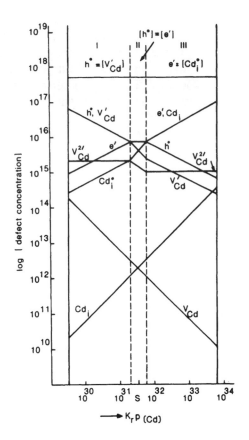

Figure 8.13 Brouwer diagram for the semiconductor compound CdTe, derived from experimentally determined values for the defect concentrations and the partial pressure of Cd vapour.

and the lines for these two defects overlap here. There is a high population of holes and the material will be a p-type semiconductor. The equally high population of single-charged cadmium vacancies will make Cd diffusion through the Cd sub-lattice easy.

At high partial pressures of Cd, in region III, we have a charge neutrality equation

$$[e'] = [Cd_i^\bullet]$$

Electrons now dominate the electronic defect population and the compound will be an n-type semiconductor. The high number of Cd interstitials suggests that Cd diffusion by interstitial or interstitialcy mechanisms would occur.

The figure shows that as the partial pressure of Cd is swept from low to high values, the compound starts out by showing p-type semiconduction which falls as the pressure rises. The material passes through a point of minimum conductivity at the stoichiometric composition, after which it becomes an n-type semiconductor with rising conductivity as the pressure increases. Thus, the electronic properties needed for the device planned can be optimized during crystal growth.

The diagram could be complemented by a similar one for Te defects. Would these be important in practice? Well that will depend on the magnitude of the equilibrium constants for the defect formation reactions compared to those utilized in drawing Figure 8.13.

8.8 Supplementary reading

The thermodynamics of solid equilibrium with gas atmospheres is considered in many textbooks of thermodynamics, and often with useful examples in textbooks of metallurgy and geology. The thermodynamics of many metal–oxygen systems are especially well characterized in view of their industrial importance. Two very useful descriptions of gas–solid equilibria are:

T.B. Reed, *Free energy of formation of binary compounds: an atlas of charts for high-temperature chemical calculations*, M.I.T. (1971).
A. Muan, The effect of oxygen pressure on phase relations in oxide systems, *Am. J. Sci.* **256**, 171–207 (1958).

A very readable account of oxide equilibrium with self-test questions is given by:

G.A. Smiernow and L. Twidwell, *J. Ed. Mod. Mater. Sci. Eng.* **1**, 223 (1979).

The thermodynamics of non-stoichiometric compounds and the relationship between thermodynamics and structures has been covered at an advanced but clear level by J.S. Anderson. See particularly:

J.S. Anderson, *Bull. Soc. Chem. France* **7**, 2203 (1969).
J.S. Anderson, *Problems of Non-stoichiometry*, ed. A. Rabenau, North–Holland, Amsterdam (1970).

The use of the phase rule is clearly explained by:

E.G. Ehlers, *The Interpretation of Geological Phase Diagrams*, Freeman, San Francisco (1972).

In this respect, the original works of J. Willard Gibbs are also extremely interesting. They have been republished in two volumes by Dover Publications (1961).

Much experimental data concerning electronic conductivity in oxides and its variation with oxygen partial pressure will be found in:

P. Kofstad, *Non-stoichiometry, Diffusion and Electrical Conductivity in Binary Metal Oxides*, Wiley–Interscience, New York (1972).

The original description of Brouwer diagrams is well worth consulting. It is:

G. Brouwer, *Philips Res. Reports* **9**, 366 (1954).

More information and a large number of examples, can be found in the following books:

F.A. Kröger, *The Chemistry of Imperfect Crystals*, 2nd Edition, North–Holland, Amsterdam (1974).
W. van Gool, *Principles of Defect Chemistry of Crystalline Solids*, Academic Press, New York (1966).

9 Point defects and planar defects

In recent years, it has been discovered that non-stoichiometric compounds use a wide and fascinating range of defect types to achieve a composition range. Many of these defects are interesting in their own right, while others are important because they occur in commercially useful materials. This chapter provides a brief tour of some of the more important of these defects and the compounds that they are particularly associated with.

9.1 Point defects in nearly stoichiometric crystals

There are a number of materials which show small but measurable departures from stoichiometry. The best known of these compounds are probably the transition metal monoxides typified by NiO and CoO, but a large number of other materials also fall into this class, for example, ZnO, CdO, Cu_2O, V_2O_3, VO_2 and NbO_2. In some of these phases the composition range spans both sides of the stoichiometric composition; NbO_2, for instance, has a reported composition range from $NbO_{1.9975}$ to $NbO_{2.003}$. In others, the materials have a composition range on only one side of the stoichiometric composition. In CoO, for example, the composition can range from approximately $Co_{0.99}O$ up to $Co_{1.00}O$ while in CdO the composition ranges from $Cd_{1.0000}O$ to approximately $Cd_{1.0005}O$.

The experimental evidence available for these compounds suggests that the composition range is due to the presence of isolated point defects. Thus, NiO and CoO, both of which possess the *NaCl* structure when fully stoichiometric, are considered to accommodate composition changes by way of a population of vacancies on the normally occupied metal positions. In the case of CdO, which also has the *NaCl* structure, the metal excess is usually considered to be due to interstitial Cd atoms or ions.

9.2 Point defect clusters

Even when the composition range of a non-stoichiometric phase remains small, more interesting defect structures can occur. In systems which contain *point defect clusters* isolated point defects have been replaced by aggregates of point defects with a well-defined structure. Two examples have been chosen to illustrate this behaviour, $\sim$FeO and UO_{2+x}.

9.2.1 Iron oxide, $\sim$FeO

Probably the best documented example of a material containing point defect clusters is $\sim$FeO, wüstite. This oxide exists over a composition range from $Fe_{0.89}O$ to $Fe_{0.96}O$ at 1300 K which broadens with increasing temperature. The original view of $\sim$FeO suggested that the compound had an *NaCl*-type structure in which the oxygen array was perfect and the non-stoichiometric composition was due to vacancies among the iron atoms, as discussed in chapter 4. However, careful structural studies have shown that isolated iron vacancies are not present at all, but instead small groups of atoms and vacancies aggregate into elements of new structure which are distributed throughout the wüstite matrix. Some of the cluster arrangements so far characterized are shown in Figure 9.1.

It is of great interest to discover that these clusters bear a strong resemblance to *fragments of the structure of Fe_3O_4*, the next higher oxide to FeO. Structurally, therefore, the oxygen excess FeO is a partly ordered assembly of fragments of the Fe_3O_4 structure distributed throughout the *NaCl* structure expected of an oxide of formula FeO. The fact that these clusters are more stable than equivalent point defect populations has been confirmed by calculation of lattice energies.

9.2.2 Uranium oxide, UO_{2+x}

Uranium oxides are of importance in the nuclear industry, and for this reason a lot of effort has been put into understanding their non-stoichiometric behaviour. The dioxide, $\sim UO_2$, crystallizes in the *fluorite* structure type with an ideal composition MX_2. For a good many years it has been known that in these compounds it is the non-metal lattice which is the seat of the non-stoichiometric variation and in the MX_{2+x} phases interstitial anions are present.

One of the earliest cluster geometries to be understood is that of the so-called *Willis 2:2:2 cluster* in the *fluorite* structure oxide UO_{2+x}. The structure of this cluster is shown in Figure 9.2. The cluster is made up from four oxygen cubes of the type shown in Figure 9.2(a) and (b) joined in the normal *fluorite* arrangement, shown in Figure 9.2(c) and (d). The cluster, shown in Figure 9.2(c) is formed by placing oxygen interstitials into the centre of the cubes at the top and bottom of the cluster, shown darker in Figure 9.2(d). Two oxygen ions which are normally at the junctions of three cubes, move into interstitial sites in the middle of the two central cubes, shown lighter in Figure 9.2(d). This creates two oxygen vacancies at these

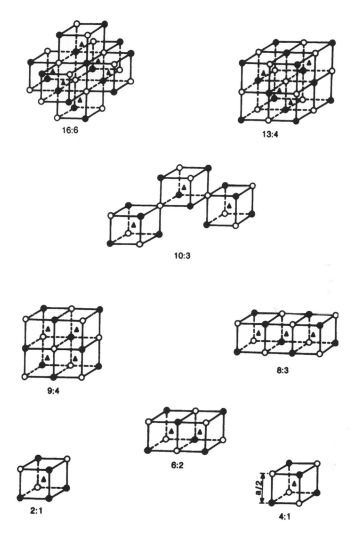

Figure 9.1 The proposed structure of clusters which are believed to occur in $Fe_{1-x}O$. The open circles represent Fe vacancies, the filled triangles show Fe in a tetrahedral coordination and the filled circles show Fe in octahedral environments. Oxygen atoms have been omitted for clarity. The numbers below each diagram represent the ratios of Fe vacancies to Fe in tetrahedral positions.

corner junctions. The cluster name comes from the fact that there are *two* normal interstitials (called $\langle 111 \rangle$ interstitials), *two* created by moving normal anions (called $\langle 110 \rangle$ interstitials) and *two* vacancies.

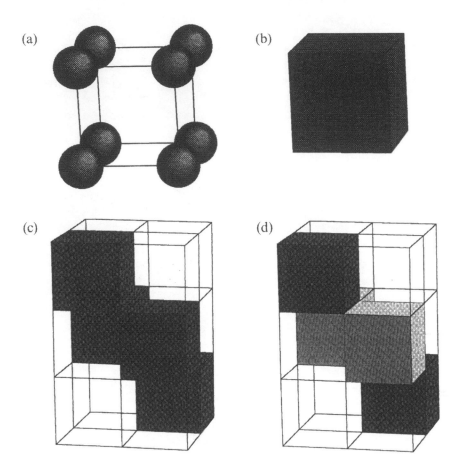

Figure 9.2   The structure of the 2:2:2 cluster in UO_{2+x}. (a) and (b) The O^{2-} cube which is part of the *fluorite* structure of UO_{2+x}. (c) The cluster, made up of four cubes. (d) Two interstitial oxygen ions centre the top and bottom cubes (darker shading) and interstitial oxygen ions from the corners, where three cubes meet, centre the middle two cubes (lighter shading).

9.3   Microdomains

As was pointed out for $\sim FeO$, point defect clusters can sometimes be recognized as fragments of another structure. At times, these fragments become large enough to be considered as a *microdomain* of the other structure that is *coherently intergrown* within the structure of the non-stoichiometric phase. In such a situation at least one of the sub-lattices of the phase must be *continuous* across both the microdomains and the parent lattice, and no changes in crystal structure should occur. This situation is different from that in which a small precipitate is formed in a structure. In this case, the precipitate will have a different crystal structure to the parent matrix and quite clear crystallographic boundaries between the two phases will be found.

A good example of microdomain formation is provided by non-stoichiometric $Ti_{1+x}S_2$. This material can be regarded as a derivative of the stoichiometric TiS_2 structure shown in chapter 4. It consists of strongly bonded TiS_2 sheets which are held together by rather weak forces. It is easy to synthesize non-stoichiometric materials with an overall composition $Ti_{1+x}S_2$. In these phases, the additional Ti atoms lie between the TiS_2 layers. It is interesting to discover that these extra atoms do not sit at random, but form microdomains with well-defined geometries. A model of the microdomain configuration found at the composition $Ti_{1.26}S_2$ is shown in Figure 9.3. In this phase, the microdomains can be regarded as tiny pieces of the TiS structure coherently embedded in the TiS_2 phase. As the amount of extra Ti goes up, the microdomains pack more closely together and change their configuration slightly.

This is rather typical of microdomain microstructures. The microdomains themselves often have a fixed composition, as does that of the matrix which contains them. The apparent non-stoichiometry of the system is caused by a

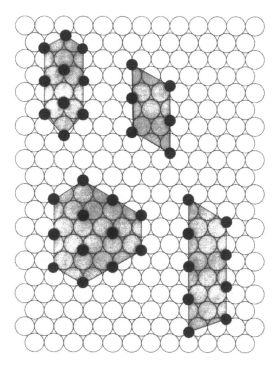

Figure 9.3 Some microdomain structures in $Ti_{1.26}S_2$. The microdomains are made up of ordered patches of Ti atoms (filled circles) lying between hexagonal layers of S atoms (open circles).

variation of the relative proportions of the two sorts of structure. Because these are both ordered variants of an identical parent structure, complete transformation from one structure to the other is possible, at least in theory.

9.4   Point defect ordering and assimilation

Defect clusters and microdomains can be regarded as precursors to structures in which all of the defects are completely ordered. In this case the defects have been *assimilated* into the structure and are, strictly speaking, no longer 'defects' but a legitimate part of the structure. One of the best known systems showing this type of behaviour is the *perovskite* phase, $SrFeO_{2.5}$. At high temperatures, this material has a cubic *perovskite* structure. The ideal composition for a *perovskite* structure would be $SrFeO_3$, and the oxygen deficit is due to disordered oxygen vacancies. At temperatures below 900 °C these oxygen vacancies order completely in a specific way. The structure formed is an example of the *brownmillerite* type, shown in Figure 9.4. In this structure (named after the mineral brownmillerite, Ca_2FeAlO_5), half of the Fe cations are in octahedral sites and half in tetrahedral sites. In this ordered state, it is better to give the phase the formula $Sr_2Fe_2O_5$ to emphasize the important structural change that has occurred. The tetrahedral sites are created by ordering the 'oxygen vacancies' in rows, as shown in Figure 9.4. This is followed by a slight displacement of the cations nearest to the vacancies so as to move them into the centre of the terahedra formed. The structure is now made up of sheets

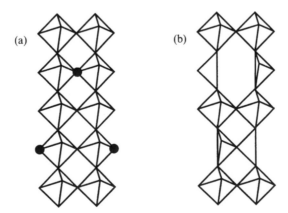

(a)   (b)

Figure 9.4 An idealized representation of the tetrahedral and octahedral framework of the *brownmillerite* structure. (a) The structure can be derived from the octahedral framework in the *perovskite* structure by the removal of oxygen ions, shown as filled circles. This generates the tetrahedra shown in brownmillerite (b). The large cations are omitted in both figures.

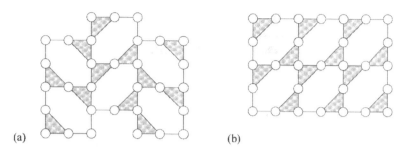

(a)   (b)

Figure 9.5 The idealized structures of (a) $Sr_2Mn_2O_5$ and $Ca_2Co_2O_5$, and (b) one of the forms of $Ca_2Mn_2O_5$. The structures are made up of corner-linked square pyramids formed by the ordered omission of oxygen atoms from the octahedra, making up the framework of the idealized *perovskite* structure. The large cations are omitted in both figures.

of *perovskite* structure interleaved with slabs of corner-linked tetrahedra reminiscent of many silicate structures.

As would be expected from the earlier chapters of this book, structures with high oxygen vacancy concentrations should show high oxygen ion diffusion. This is important in *selective-oxidation catalysts*. These materials are used to oxidize hydrocarbons at specific sites to produce precise reaction products. An important requisite of these compounds is that when necessary they take up oxygen from the gas phase rapidly and the catalytic reaction can be sustained at other times by rapid diffusion of oxide ions from the crystal bulk.

There are a number of non-stoichiometric, selective-oxidation catalysts which contain high concentrations of ordered oxygen vacancies. To illustrate the types of ordering encountered the three phases $Ca_2Mn_2O_5$, $Sr_2Mn_2O_5$ and $Ca_2Co_2O_5$ are shown in Figure 9.5. They all have the *brownmillerite* composition and are derived from a parent *perovskite* structure, ABO_3, but use a different way of ordering the oxygen vacancies. In fact, the oxygen vacancies are ordered so as to generate linked square pyramids, as shown in the figure.

The examples given above reveal one important feature involved in the incorporation of the oxygen vacancies into the structure. At least some of the cations involved must be able to take up a variety of different coordination polyhedra. We will see in the following chapter that this is an important feature of the copper containing high temperature super-conductors.

9.5 Interpolation

Interpolation in non-stoichiometric compounds was introduced in chapter 4. The tungsten bronzes, which have a formula M_xWO_3, are interesting examples of these phases because they show metallic properties and vivid colours. In chapter 4, the cubic *perovskite* tungsten bronzes typified by Li_xWO_3 were described. In addition, two other tungsten bronze types exist, shown in Figure 9.6. These are both made up of corner-linked WO_6 octahedra, arranged to form pentagonal and square tunnels in the *tetragonal tungsten bronze structure*, or hexagonal tunnels in the *hexagonal tungsten bronze structure*. Variable filling of the tunnels by metal atoms gives rise to wide stoichiometry ranges. In the sodium tetragonal tungsten bronze phases, both the pentagonal and square tunnels are partly filled. (The bronze Na_xWO_3 has the tetragonal tungsten bronze structure for values of x between 0.26 and 0.38. Outside this range, Na_xWO_3 adopts the cubic *perovskite* tungsten bronze structure.)

However, in many tetragonal tungsten bronzes, typified by Pb_xWO_3 and Sn_xWO_3, in which x can take values from approximately 0.16 to 0.26, only the pentagonal tunnels are occupied. In the hexagonal tungsten bronzes, it is

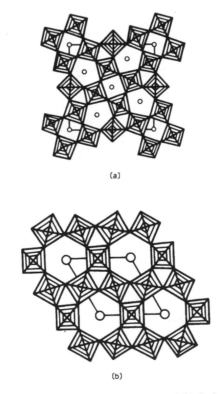

(a)

(b)

Figure 9.6 (a) The tetragonal tungsten bronze structure and (b) the hexagonal tungsten bronze structure. The shaded squares represent WO_6 octahedra, which are linked to form pentagonal, square and hexagonal tunnels. These are able to contain a variable population of metal atoms, shown as open circles.

the hexagonal tunnels which are partly occupied. The composition of the hexagonal tungsten bronze phase K_xWO_3, extends from $x = 0.19$ to 0.33 in this way.

The *hollandite* structure is another arrangement containing rather wide tunnels, as can be seen from Figure 9.7. $Ba_2Mn_8O_{16}$ is a typical example of these phases. The tunnels can be filled with large metal atoms in variable proportions to create non-stoichiometric *hollandites*. *Hollandites* formed from the parent compound TiO_2, have received considerable attention in recent years for possible use in the storage of radioactive materials. These radioactive wastes must be turned into a stable solid and then stored for extended periods. The reason for choosing TiO_2 is that it is a naturally occurring mineral which has shown itself to be stable over geological time scales. Mixtures of TiO_2, when heated with large radioactive metal ions, form *hollandites* in which the radioactive cations are trapped in the tunnels. The *hollandite* matrix is an inert shell which would then need to be encapsulated before storage. Although this technology has not yet become commercially viable the idea of using tunnel compounds for the isolation of dangerous ions is still being actively studied experimentally.

9.6 Planar faults and boundaries

Instead of using defects confined to one or a few atom sites, some systems utilize *planar faults* or *defects* (sometimes called *extended defects*) to change the composition. A planar fault in a crystal is created by schematically cutting the crystal into two and rejoining the pieces. The three ways in which this can occur are shown in Figure 9.8. In Figure 9.8(a) the composition of

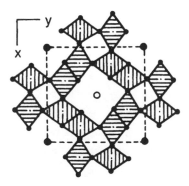

Figure 9.7 The *hollandite* structure. The framework is composed of continuous chains of metal–oxygen octahedra, shown as shaded diamonds, which are linked so as to form square tunnels. A variable population of cations, shown as filled or open circles, can occupy these tunnels to give the materials significant composition ranges.

the crystal has not changed as the fault has only slipped one part of the crystal past the other. These are called *anti-phase* boundaries. In Figure 9.8(b) a plane of atoms has been removed at the fault. In this case the composition of the crystal has changed slightly. These boundaries are called *crystallographic shear planes*. The fault shown in Figure 9.8(c) is a mirror plane, and is called a *twin plane*. Some twin planes do not change the composition of the crystal while at others atoms are lost and a composition change can result. If faults which alter the composition are introduced in variable numbers and distributed at random, we will generate a crystal with a variable composition and a non-stoichiometric phase is produced. New coordination polyhedra are often created in the vicinity of the fault that are

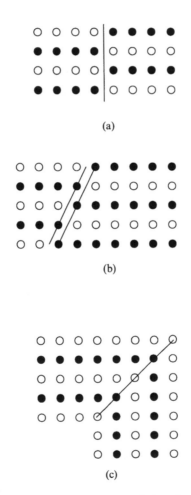

(a)

(b)

(c)

Figure 9.8 Planar boundaries in crystals. (a) An anti-phase boundary. (b) A crystallographic shear plane. (c) A twin plane.

not present in the parent structure. These may provide sites for novel chemical reactions or introduce changes in the physical properties of significance compared to those of the parent structure.

If planar boundaries are introduced into a crystal in an ordered way a new and longer unit cell will result. Crystallographically, a *new phase* will have formed. Strictly speaking the faults are no longer defects and in this case are simply referred to as *boundaries*. If phases with different boundary spacings are synthesized, a *homologous series* of new phases will be generated. Each phase will be characterized by the separation between the ordered planar boundaries. In addition, each phase will have a fixed composition, although the formula may involve large integers. When the boundaries introduce a composition change, like those in Figure 9.8(b), each member of the homologous series will differ in composition from its neighbours by a small but definite amount.

9.7   Crystallographic shear phases

The phenomenon of crystallographic shear (*CS*) seems to be important mainly in the transition metal oxides WO_3, MoO_3 and TiO_2, and provides a mechanism for altering the anion to cation ratio in these materials without either changing the shape of the anion coordination polyhedra of the metal atoms or introducing point defects.

9.7.1   Crystallographic shear in tungsten oxides

The simplest compounds that we can use to illustrate this process occur in the tungsten–oxygen system. If tungsten trioxide is very slightly reduced (to approximately $WO_{2.998}$), random {102} *CS* planes form in the crystal in order to accommodate the oxygen loss, as can be seen in Figure 9.9. The structure of the parent oxide, WO_3, is shown in Figure 9.10(a). To form a {102} *CS* plane, imagine that a sheet of oxygen atoms lying on a (210) plane has been removed from the crystal and the two sides rejoined. The structure of the resulting {102} *CS* plane is shown in Figure 9.10(b). It consists of blocks of four-edge-shared octahedra in a normal WO_3-like matrix. One oxygen ion is lost per block of four edge-shared octahedra.

Although slight reduction produces disordered {102} *CS* planes, as the composition approaches $WO_{2.95}$ these become ordered to form a homologous series. The composition of any phase is given by W_nO_{3n-1}, where n represents the number of octahedra separating the *CS* planes, measured along the direction shown by an arrow in Figure 9.10(b). These oxides are also known as *crystallographic shear phases*. The lower limit of the {102} *CS* series is approximately $W_{18}O_{53}$.

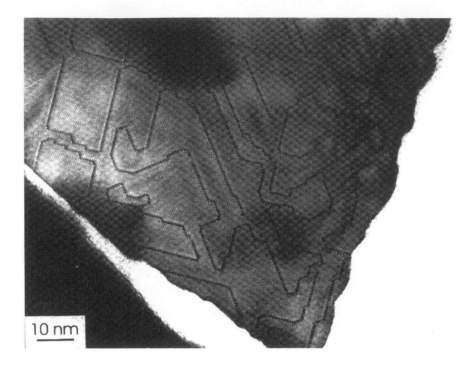

Figure 9.9 An electron micrograph showing random {102} CS planes in a crystal of slightly reduced WO_3.

If reduction continues something remarkable happens. When the composition falls to about $WO_{2.94}$, {102} CS planes are replaced by {103} CS planes. The structure of these CS planes consists of blocks of six edge-shared octahedra, as shown in Figure 9.10(c). In a {103} CS plane, two oxygen ions are lost per block of six edge-shared octahedra. These {103} CS planes are usually fairly well ordered and give rise to a homologous series of oxides with a general formula of W_nO_{3n-2} over a composition range of approximately $WO_{2.93}$ to $WO_{2.87}$, that is from $W_{25}O_{73}$ to $W_{16}O_{46}$. An electron micrograph of a well-ordered {103} CS structure is shown in Figure 9.11.

Why does the change of CS plane occur? In fact it is a response to minimize lattice strain. The cations in the blocks of edge-shared octahedra within the CS planes repel each other and so set up strain in the lattice. The total lattice strain depends on the number of CS planes present. When the composition of a reduced crystal reaches about $WO_{2.92}$ the {102} CS planes are very close and the lattice strain becomes too great for further oxygen loss to take place. However, if the CS plane changes to {103} the same degree of reduction is achieved with half the number of CS planes and so the total lattice strain falls. This is because each block of six edge-shared octahedra

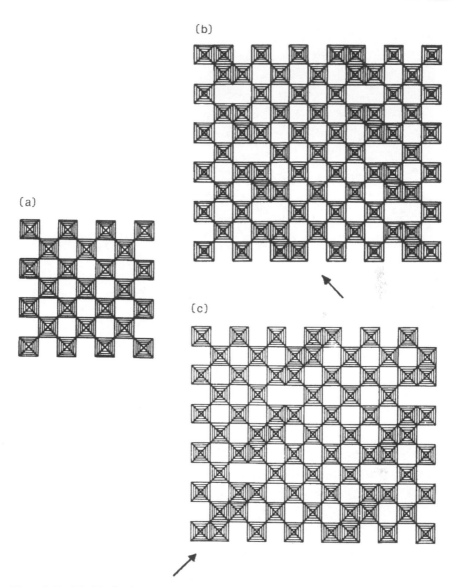

Figure 9.10 The idealized structures of (a) WO_3; (b) $\{102\}$ CS planes; and (c) $\{103\}$ CS planes. The shaded squares represent WO_6 octahedra which are linked by corner-sharing in WO_3 and by edge-sharing within the CS planes. The arrows in (b) and (c) show the direction in which octahedra are counted in order to measure the value of n in the formulae W_nO_{3n-1} and W_nO_{3n-2} for the $\{102\}$ and $\{103\}$ series, respectively.

loses two oxygen ions compared to one for a block of four edge-shared octahedra. The decrease in lattice strain is so great that further reduction, now using $\{103\}$ CS planes, can take place.

Figure 9.11   Electron micrograph of a {103} *CS* phase. The dark dashes are the blocks of six edge-sharing octahedra which make up the *CS* planes. Some defects in the ordering of the *CS* plane are also apparent. The spacing of the *CS* planes reveal that the material has a composition close to $W_{18}O_{52}$.

9.7.2   Crystallographic shear in titanium oxides

When the rutile form of titanium dioxide, shown in Figure 9.12, is reduced crystallographic shear planes are used to accommodate the oxygen loss. In fact, the *CS* behaviour of TiO_{2-x} is remarkably similar to that encountered in WO_{3-x}. For example, small degrees of reduction are accommodated on {132} *CS* planes. Initial reduction produces random *CS* planes, while increased reduction leads to a homologous series of *CS* phase with a series formula Ti_nO_{2n-1} and n taking values from approximately 32 down to about 16. The structure of this *CS* plane is shown in Figure 9.13(a). At greater degrees of reduction {132} *CS* planes change over to {121} *CS* planes, shown in Figure 9.13(b). The change-over occurs at compositions between approximately $TiO_{1.93}$ and $TiO_{1.90}$. The {121} series also has a formula Ti_nO_{2n-1}, and the oxides run from Ti_4O_7 to Ti_9O_{17}.

In these figures, the anion packing has been emphasized rather than the metal coordination polyhedra to illustrate the distinction between the organization of a *CS* phase and the occurrence of point defects. In chapter 4 the relationship between density and defect structure was discussed. Measurement of the density of reduced TiO_{2-x} suggests that interstitial Ti ions form. Examination of the structures in Figure 9.13 shows that extra Ti ions are present in the crystal compared to TiO_2 (because oxygen has been

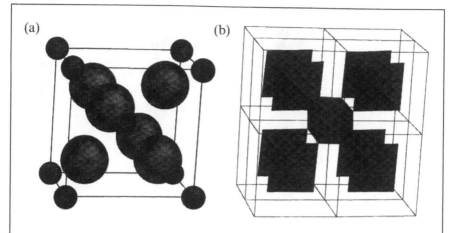

Figure 9.12 (a) The rutile structure of TiO_2 shown in a perspective view and (b) emphasizing the TiO_6 octahedra.

The rutile (TiO_2) structure

Titanium dioxide crystallizes in several forms. The most important is the rutile form. The unit cell is tetragonal with $a = 0.4594$ nm, $c = 0.2958$ nm. There are two formula units in a unit cell.

Other materials which crystallize in the *rutile* structure include SnO_2, MgF_2 and ZnF_2. A number of oxides which show metallic or metal–insulator transitions, for example VO_2, NbO_2, CrO_2, all have a slightly distorted form of the structure.

lost) and they occupy interstitial positions. However, these 'interstitials' are integrated into the structure, with the consequence that interstitial 'point defects' are not present.

9.7.3 Impurities and CS in WO₃ and TiO₂

All the systems which rely on *CS* to accommodate oxygen loss reveal amazing degrees of complexity. This is especially apparent when 'oxygen loss' is achieved by reaction with other oxides. If, for example, WO_3 is reacted with small amounts of Nb_2O_5, each pair of Nb ions which enter the WO_3 lattice require a reduction in the amount of oxygen ions present by

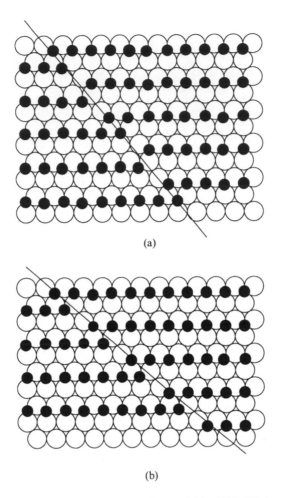

(a)

(b)

Figure 9.13 The structures of (a) a {132} *CS* plane and (b) a {121} *CS* plane in TiO$_2$, drawn so as to emphasize the packing of anions, shown as open circles, and cations, shown as filled circles.

one. Oxygen vacancies do not form and the problem is solved by the use of *CS* planes. These are not on {102} or {103}, as one would expect from pure WO$_3$, but form initially on {104} planes, which change to {001} planes as the Nb concentration increases. At compositions between these two regions spectacular wavy *CS* 'planes' are found, as shown in Figure 9.14. These do not form in the binary tungsten–oxygen system and reveal that complex factors are involved in determining the planes on which *CS* actually occurs.

This conclusion is substantiated by an example from the TiO$_2$ system. One of the most remarkable *CS* systems is produced if TiO$_2$ is reacted with Cr$_2$O$_3$. Small amounts of Cr$_2$O$_3$ cause {132} *CS* planes to form and larger

Figure 9.14 An electron micrograph showing wave-like *CS* planes in WO_3 reacted with Nb_2O_5.

amounts cause $\{121\}$ *CS* planes to form, just as in the TiO_{2-x} itself. However, when the overall composition of the oxide lies between approximately $(Ti,Cr)O_{1.93}$ and $(Ti,Cr)O_{1.90}$ a *swinging CS region* occurs. What this means is that the *CS* plane indices gradually transform from $\{132\}$ towards $\{121\}$ as the composition range is traversed. What is amazing, though, is that the *CS* planes in these intermediate compositions are always *perfectly ordered*. This should be stressed. Every composition prepared has an ordered arrangement of *CS* planes, with a definite spacing and *CS* orientation. Within this composition range, any composition at all seems to have a unique ordered *CS* structure.

9.8 Chemical twinning

9.8.1 Lead–bismuth sulphosalts

The use of *chemical twinning* (*CT*) to accommodate non-stoichiometry using twin planes can be illustrated by referring to the PbS–Bi_2S_3 system. Lead sulphide, PbS, often referred to by its mineral name, *galena*, has the *rock salt* structure. This is shown in projection down the [110] direction rather than the normal [100] direction of earlier diagrams, in Figure 9.15(a). From a crystal chemistry viewpoint, the problem is how to accommodate Bi_2S_3 into

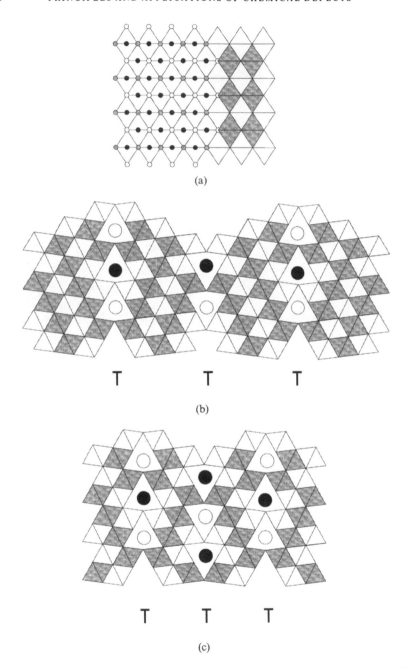

Figure 9.15 The structures of (a) PbS projected onto (110); (b) lillianite projected onto (001) and (c) heyrovskyite projected into (001). In each diagram the structures are shown as a packing of metal–sulphur octahedra; those at a higher level being shown in light relief and those at a lower level are darker. The structures (b) and (c) can be regarded as made up of slabs of PbS structure joined along twin planes, marked T, which contain metal atoms in trigonal prismatic coordination.

galena. Two chemical problems arise, the Bi^{3+} will not fit into octahedra like the Pb and for each pair of Bi ions added to the galena three extra sulphur ions must be fitted in as well.

The solution is given by looking at the structures of the two phases, *heyrovskyite*, $Pb_{24}Bi_8S_{36}$, and *lillianite*, $Pb_{12}Bi_8S_{24}$, shown in Figure 9.15. The structures of these minerals, which belong to a group of mineral *sulphosalts*, are made up of twinned strips of PbS. This is rather clever. The twinning of the galena structure has achieved the two objects needed. First, the extra sulphur ions have been incorporated into the structure without the use of interstitials. Secondly, the twin planes introduce new and larger sites which do not exist in the original PbS structure. The Bi ions occupy these quite comfortably. The twinned nature of the phase heyrovskyite is clearly revealed in the electron micrograph shown in Figure 9.16.

Figure 9.15 shows that the width of the strips of PbS is different in the two structures lillianite and heyrovskyite. They are members of a homologous series. Other members of the series will form as the PbS regions take on other thicknesses. Many such phases have been found, particularly in mineral samples. Some of these phases use an interesting structural method of fitting intermediate compositions into the series. Alternating slabs of galena which are of different widths are used. It is certain that continued examination of these sulphides will produce other surprises.

Figure 9.16   An electron micrograph of the chemically twinned phase heyrovskyite, which is built up of twinned galena-like slabs, seven PbS octahedra in width. Careful measurement reveals that some slabs are wider than normal and contain eight octahedra.

9.8.2   Molybdenum oxides and phosphate bronzes

Chemical twinning also occurs in a number of oxide systems. One rather elegant example is provided by the orthorhombic form of the oxide Mo_4O_{11}. This oxide is made up of twinned slabs of a structure similar to that of WO_3 (but with a composition of MoO_3), linked by MoO_4 tetrahedra as shown in Figure 9.17. There are six MoO_6 octahedra in each twinned slab and the structure is the $n = 6$ member of the homologous series $Mo_{n+2}O_{3n+4}$, Mo_8O_{22}. The $n = 7$ phase, with seven octahedra in each slab, $Mo_{7.8}W_{1.4}O_{25}$ is also known.

An extensive series of structures with the same basic twinning motif are formed when WO_3 is reacted with the acidic PO_4 group. The formula of the materials is $P_4O_8(WO_3)_{2n}$ and they are referred to as *phosphate bronzes*. The PO_4 groups replace the MoO_4 groups found in the molybdenum oxide and link WO_3-like slabs to produce compounds which are structurally equivalent to the molybdenum oxides. However, it is interesting to discover that a wider range of n values occurs.

9.9   Planar intergrowths

In *CS* and *CT* phases, the structures on each side of the planar boundary are the same. It is quite easy to conceive of the situation where structures on

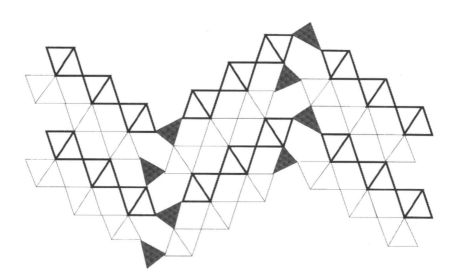

Figure 9.17 The idealized structure of Mo_4O_{11} (orthorhombic) which is identical to the structure of the $n = 6$ member of the phosphate bronze phases $P_4O_8(WO_3)_{2n}$. The diamond shapes represent MoO_6 or WO_6 octahedra and the shaded triangles are MoO_4 or PO_4 tetrahedra.

each side of the boundary are different, in which case we meet with the phenomenon of *intergrowth*. Here we will mention just two examples of intergrowth systems which show the amazing complexity of non-stoichiometric compounds.

9.9.1   Perovskite-related structures in the $Ca_4Nb_4O_{14}$–$NaNbO_3$ system

An extensive intergrowth series is found in the system bounded by the end members $Ca_4Nb_4O_{14}$ and $NaNbO_3$. The structure of $NaNbO_3$ is the same as that of the mineral *perovskite*, $CaTiO_3$. In $Ca_4Nb_4O_{14}$ the structure is seen to be composed of slabs of *perovskite* type structure four octahedra in thickness, as shown in Figure 9.18. In this material, the slabs of *perovskite* structure are united by lamellae of composition CaO which have the *rock salt* structure and the oxides can be thought of as intergrowths of *perovskite* with *rock salt*. The principle phases which form in this system can be described by the series formula $(Na,Ca)_nNb_nO_{3n+2}$, where n represents the number of metal–oxygen octahedra in each slab. Thus, $Ca_4Nb_4O_{14}$ is the $n = 4$ member of the series and $NaNbO_3$ is the $n = \infty$ member.

Although the structures just described are interesting, they do not even begin to approach the complexity that lies in the system. Between just the $n = 4$ and $n = 5$ phases *several hundred* other structures have been

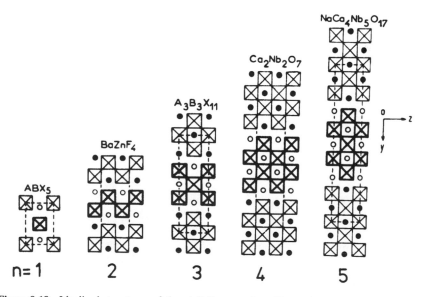

Figure 9.18   Idealized structures of the $A_nB_nO_{3n+2}$ series of layered perovskite related phases for values of n ranging from 1 to 5. The $n = 4$ structure is represented by $Ca_4Nb_4O_{14}$. Infinite slabs make up the *perovskite* structure of $NaNbO_3$.

characterized. These are made up of different arrangements of $n = 4$ and $n = 5$ units. As an example, the composition $Na_{0.5}Ca_4Nb_{4.5}O_{15.5}$ is the formula of an oxide with one unit each of the $n = 4$ phase $Ca_4Nb_4O_{14}$ and the $n = 5$ phase $NaCa_4Nb_5O_{17}$. It could be given a series formula of $n = 4.5$. The structure consists of alternating $n = 4$ and $n = 5$ lamellae in a perfectly ordered arrangement. Between compositions of the $n = 4.5$ phase, with an oxygen to metal ratio of $MO_{1.7222}$ and $MO_{1.735}$, about 50 structures have been characterized. This is remarkable!

Similar complexity exists between the other members of the series shown in Figure 9.18, so that over the whole of the phase region an enormous number of structures can be prepared. It is not surprising that research on these materials proved difficult!

9.9.2 Intergrowth tungsten bronzes

The intergrowth tungsten bronzes are intergrowths formed by the parent phases WO_3 and the hexagonal tungsten bronze structure, which have been illustrated in Figures 9.6 and 9.10. The hexagonal tungsten bronze structure is formed when potassium is reacted with WO_3, within the composition range from $K_{0.19}WO_3$ to $K_{0.33}WO_3$. What happens when the amount of potassium is less than the minimum needed to form the hexagonal tungsten bronze structure? In this case an intergrowth between the hexagonal tungsten bronze structure and WO_3 is found. Usually the strips of hexagonal structure are two tunnels in width in the K_xWO_3 system, and this structure is illustrated in Figure 9.19(a). Similar intergrowth tungsten bronzes are also known in a number of other systems, including Rb_xWO_3, Cs_xWO_3, Ba_xWO_3, Sn_xWO_3 and Pb_xWO_3. In the non-alkali metal intergrowth bronzes, single tunnels seem to be preferred. This structure is shown in Figure 9.19(b) and an electron micrograph of the single tunnel Ba_xWO_3 intergrowth bronze is shown in Figure 9.20.

In all these phases, the separation of the strips of hexagonal tunnels increases as the concentration of the large interpolated atoms decreases, and a homologous series of compounds form. Because the hexagonal tungsten bronze parent structure is able to tolerate a considerable range of composition due to variable filling of the hexagonal tunnels, these intergrowth bronzes behave in the same way. In these phases, therefore, we have two ways of accommodating the change of composition, either by changing the relative numbers of hexagonal tunnels with respect to the WO_3 matrix, or else by varying the degree of filling of the hexagonal tunnels themselves.

(a)

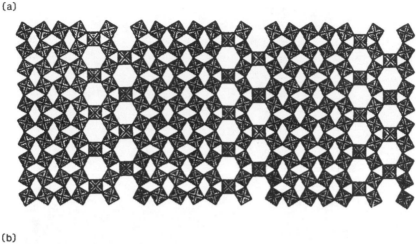

(b)

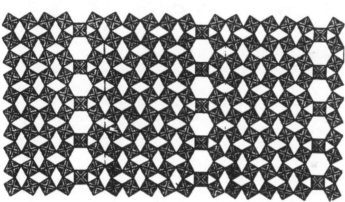

Figure 9.19 The idealized structures of two intergrowth tungsten bronze phases (a) containing double rows of hexagonal tunnels, and (b) containing single rows of tunnels. The tungsten trioxide matrix is shown as shaded squares and the hexagonal tunnels are shown empty, although in the known intergrowth tungsten bronzes the tunnels contain variable amounts of metal atoms.

9.10 Case study: the Nb_2O_5 block structures

A remarkable group of niobium oxides which have compositions close to Nb_2O_5 have been studied for many years. Early studies (complicated for the reader by the fact that the element is called Columbium in some reports) suggested that these materials had extensive non-stoichiometric composition ranges. It is convenient to start with the oxides which lie between the composition limits of $NbO_{2.5}$ and $NbO_{2.40}$. This was considered to consist of a single phase region due to point defects in the parent Nb_2O_5 structure. Similarly when Nb_2O_5 was reacted with other oxides extended composition

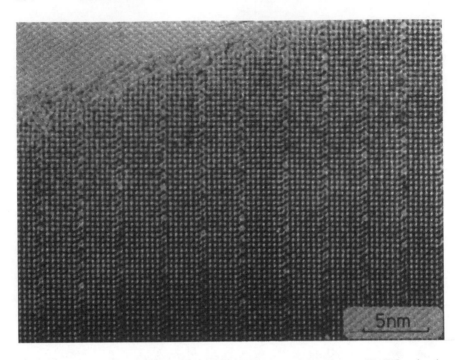

Figure 9.20 An electron micrograph of the intergrowth tungsten bronze Ba_xWO_3, showing single rows of hexagonal tunnels. The square arrays of black spots represent the tungsten ions in the WO_3 portions of the structure. Some of the tunnels are seem to be empty or only partly filled with barium.

ranges were attributed to the resulting compounds. For example, it was believed that Nb_2O_5 could incorporate up to 50 mole % WO_3 into its structure by using point defects. The composition range of the material was thought to extend from $NbO_{2.5}$ to $(Nb,W)O_{2.7}$.

The breakthrough came when the structure of $H\text{-}Nb_2O_5$, which is the stable form of the oxide at high temperatures, was determined. It is shown in Figure 9.21(a). As can be seen, it is composed of columns of material with a WO_3-like structure formed by two intersecting sets of CS planes. These columns, in projection, look like rectangular blocks, and hence a common name for these materials is *block structures*. The blocks have two sizes (3×4) octahedra and (3×5) octahedra, neatly fitted together to form the structure.

CS planes are associated with oxygen loss and that is so in this case too, even though Nb_2O_5 is not a *reduced* oxide. The fact is that the Nb^{5+} ion strongly prefers to sit in an oxygen-coordination polyhedron which is octahedral. If the octahedra are linked by corners the composition would be NbO_3, which is not possible because the highest valence Nb can support is

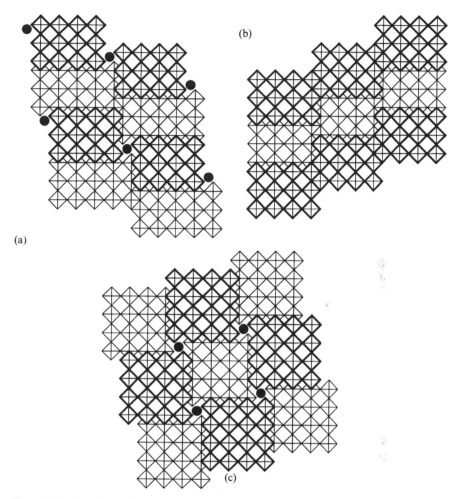

Figure 9.21 The idealized structures of some niobium oxide block structures. (a) H-Nb_2O_5; (b) $Nb_{12}O_{29}$; and (c) $W_3Nb_{14}O_{44}$. The squares represent MO_6 octahedra which are corner-sharing at the column centres and edge-sharing at the periphery of each column. The filled circles represent metal atoms in tetrahedral coordination at block junctions.

$5+$. However, in Nb_2O_5 the Nb ions can retain octahedral coordination by the use of *CS* to eliminate some oxygen. The centres of the blocks have a composition NbO_3 and the block boundaries have a composition NbO_2.

When H-Nb_2O_5 is reduced, a series of new block structures form. The non-stoichiometric composition range is actually occupied by a series of different block structures. These have smaller block sizes, which tells us that the amount of the 'NbO_3' part is decreasing and the amount of 'NbO_2' part is increasing. The most reduced oxide, $Nb_{12}O_{29}$, with blocks of (3×4) octahedra in size, is shown in Figure 9.21(b). Between this oxide and

H-Nb_2O_5, the oxides $Nb_{22}O_{54}$, $Nb_{47}O_{116}$, $Nb_{25}O_{62}$, $Nb_{39}O_{97}$ and $Nb_{53}O_{132}$ are found. Despite the formulae, which were referred to as 'grotesque' by the crystallographer A.D. Wadsley, who first unravelled these phases, all the structures are simple block jigsaws, neatly fitted together into a coherent structure.

The same thing happens when H-Nb_2O_5 reacts with many other oxides. To return to the case of WO_3, the composition needs to approach MO_3 and larger block sizes are needed. A considerable number of these span the originally reported non-stoichiometric composition range. They are typified by $W_3Nb_{14}O_{44}$, shown in Figure 9.21(c), made up of (4 × 4) blocks.

Although it might be thought that reactions between these phases, or the interconversion of one block size to another, would be difficult, the reverse is true and the reactions of the block structures take place very rapidly. For example, the reaction between Nb_2O_5 and WO_3 produces perfectly ordered phases within 15 min at 1400 K and a variety of more disordered non-stoichiometric phases in shorter times. An example of a partly ordered structure is shown in Figure 9.22.

9.11 Pentagonal column phases

Another group of compounds which contain three-dimensional faults are the so-called *PC* structures. These are a fairly large group of metal oxides which have, as their basic structural motif, the *pentagonal column*. This consists of a pentagonal ring of five MO_6 octahedra; the tunnel formed being filled with an alternating chain of oxygen and metal atoms to form a pentagonal column, as shown in Figure 9.23(a). It seems, at first sight, unlikely that such a unit could fit readily into a host structure, but nevertheless it is found that groups of *PC*s can coherently exist within a WO_3 type of matrix. As a number of oxides can adopt this structure or a distorted version of it, a wide range of *PC* phases can form.

As in previous discussions, two alternatives exist. If the *PC* elements are perfectly ordered then we generate one or more structurally related homologous series of ordered phases. Some examples of these phases are given by the oxide Mo_5O_{14} illustrated in Figure 9.23(b), the tetragonal tungsten bronze structure illustrated in Figure 9.6(a) and the oxide $Nb_{16}W_{18}O_{64}$, which is an ordered variant of the tetragonal tungsten bronze structure in which only a percentage of the available tunnel sites are filled, shown in Figure 9.23(c). If the *PC* elements in the host structure are disordered a non-stoichiometric compound is generated. Such disorder occurs, for example, when WO_3 is reacted with Nb_2O_5 for short periods of time at temperatures below about 1500 K, as the electron micrograph in Figure 9.24 shows.

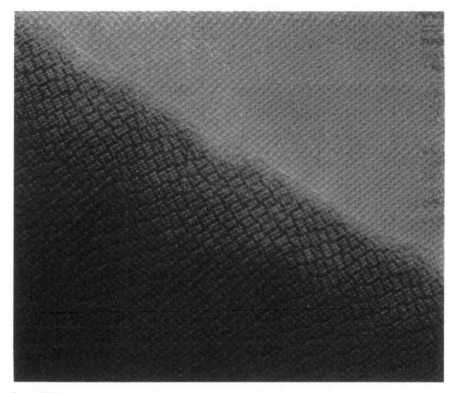

Figure 9.22 An electron micrograph of a disordered block structure formed by reacting Nb_2O_5 with WO_3. The image clearly shows the block outlines. A number of different block sizes can be seen, each of which will correspond to a distinct composition.

In all of the PC phases, another form of non-stoichiometric variation is possible. In this case the O–M–O chains which occupy the pentagonal tunnels can be incomplete, or else extra atoms can occupy tunnels which are normally empty. Interpolation is, therefore, also possible in these phases, thus matching the behaviour of the intergrowth tungsten bronzes discussed earlier.

9.12 Defect-free structures: modulated and incommensurate phases

9.12.1 Vernier structures

These structures make use of a novel way to accommodate variations in the anion to cation ratio which does not rely on point defects at all. It is possible to show how this can come about in a non-stoichiometric material by considering the orthorhombic phases formed when the oxyfluoride, YOF,

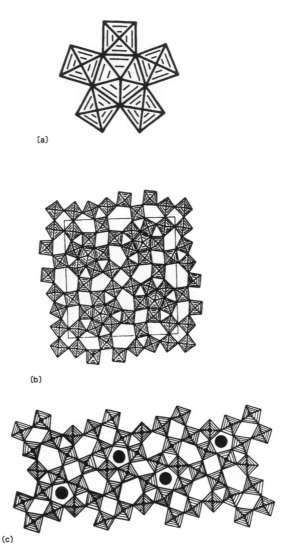

Figure 9.23 Some structures containing pentagonal columns: (a) an isolated pentagonal column; (b) the Mo_5O_{14} structure; and (c) the ordered tetragonal tungsten bronze structure of $Nb_{16}W_{18}O_{64}$, in which the square tunnels and some of the pentagonal tunnels are empty. The filled tunnels are indicated by filled circles.

reacts with small amounts of YF_3. The phases form in the composition range between $YX_{2.130}$ and $YX_{2.220}$. where X represents the anions (O, F).

The complexity of this system was unravelled by way of careful X-ray diffraction work. On powder photographs, the strong reflections correspond to that of the *fluorite*-type cell possessed by YOF. However, numerous faint lines can also be seen on these films indicating that the system is structurally

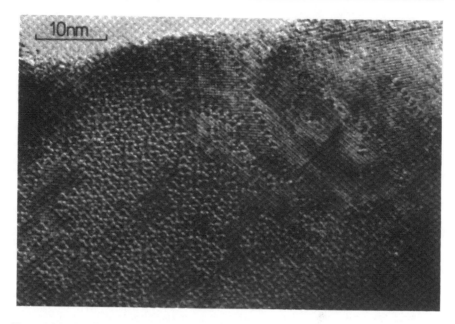

Figure 9.24 An electron micrograph of a disordered *PC* phase showing pentagonal columns, which are imaged as pairs of black areas separated by a line of white contrast, distributed at random in a WO_3 matrix.

complex. The positions of the lines change almost imperceptibly on moving from one composition to another. Moreover, the lattice parameter of the ZrOF cell appears to change smoothly as the composition varies, in agreement with Vegard's law. At first sight the system seems to be uncomplicated and a description of the non-stoichiometry in terms of cation vacancies or anion interstitials seems reasonable. However, a very careful interpretation of the X-ray results, taking into account the faint lines, yields a much more interesting picture of the real structural complexity present. It has been found that within the non-stoichiometric region every composition prepared has a different structure, and a large number of different ordered phases exist forming a homologous series of compounds with a general formula $Y_nO_{n-1}F_{n+2}$.

The structures of these phases can be thought of as made up of unit cells of YOF stacked up in a sequence which is sometimes interrupted in an ordered way allowing extra anions to be incorporated into the structure. The nature of the interruptions responsible for this is remarkable. In the *fluorite* structure the anions all lie on the corners of cubes and so form *square anion nets* in projection, as shown by the continuous lines in Figure 9.25. In the non-stoichiometric phases some of these square anion nets are changed from square into *hexagonal anion nets*, composed of triangles shown as dotted lines in Figure 9.25. Now these latter nets contain a *higher density* of anions

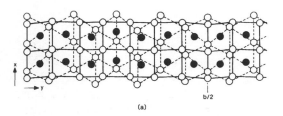

(a)

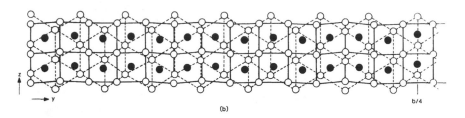

(b)

Figure 9.25 The structures of the vernier phases (a) $Y_7O_6F_9$ and (b) $Zr_{108}N_{98}F_{138}$ projected along [100]. The anion nets at $x/a = 0$ are shown as broken lines and those at $x/a = 1/2$ as full lines. The cations are shown as filled circles. If the projection of the anion nets is considered it is seen that a vernier relationship exists between the upper and lower layers.

than the square nets, and this allows the extra atoms to fit readily into the structure. The outcome is illustrated in Figure 9.25 for the phase, $Y_7O_6F_9$, which is the $n = 7$ member of the homologous series of phases $Y_nO_{n-1}F_{n+2}$, and the related compound $Zr_{108}N_{98}F_{138}$. The name *vernier structure* can now be appreciated. The positions of the metal atoms in this compound are the same as those in the *fluorite* structure parent. However, the two sorts of anion arrays are in a vernier relationship to one another. This means that a whole number, N, of squares will fit exactly with the whole number of triangles M. In Figure 9.25(a), $Y_7O_6F_9$, it is seen that seven squares fit with eight triangles.

Vernier structures are found in a number of other systems which have structures based on the *fluorite* type. The structures of these non-stoichiometric phases evolve in a continuous fashion. The pitch of the vernier will vary from one compound to another in order to accommodate the correct anion to cation proportions in the structure. You can understand this by imagining that the hexagonal anion net in Figure 9.25 shrinks or expands slightly. The match between the two structures will now occur at different values of M and N. A vernier in which 11 squares fits with 13 triangles is found in $Zr_{108}N_{98}F_{138}$ shown in Figure 9.25(b). Just a small change in the anion to cation ratio would make the vernier fit with five squares and six triangles or six squares and seven triangles. One can see that this rather subtle process allows for all of the compositions to be fitted in to an ordered structure with no defects present.

9.12.2 Infinitely adaptive compounds

In the last few sections of this chapter we have come across instances of considerable structural complexity occurring within fairly narrow stoichiometry ranges, in which, for the systems involved, any non-stoichiometric composition can be accommodated by an ordered structure. Moreover, in many systems the same composition can be accommodated by many structures. Thus, in the TiO_2–Cr_2O_3 swinging CS phases any composition can be achieved by a variety of CS plane spacings lying along different directions. In the block structures any one composition can be made from different block sizes and linkages. This a very remarkable state of affairs, not considered to be possible until recently. The solid state chemist J.S. Anderson coined the singularly apt name *infinitely adaptive compounds* for these and other phases which fall into this class.

9.12.3 Modulated or incommensurate structures

Vernier structures and some other infinitely adaptive compounds are increasingly being referred to as *modulated structures* or *incommensurate structures*. Modulated structures or incommensurate structures (there is still no consensus on a definite name for these materials) are described in terms of a fairly simple unit cell such as the *fluorite* type cell. This is often referred to as the *sub-cell* of the structure. The structure is described in terms of a *modulation wave* which has a slightly different wavelength than the sub-cell, that is, it is *incommensurate* with the sub-cell. The purpose of the modulation wave is to describe how the structure deviates from that of the sub-cell itself. If, for example, some of the atoms in a *fluorite* type sub-cell to not occupy the normal positions but positions described by an incommensurate modulation wave, a *positionally modulated* structure is formed. Alternatively, the modulation wave might describe a distribution of vacant sites with respect to the sub-cell, found in a *compositionally modulated* structure.

To give a concrete example, let us look at the deceptively simple oxide Ta_2O_5. It has been known for a long time that the Ta atoms in this phase form a uniform hexagonal array. The difficulty rests in describing the positions of the oxygen atoms. As with the vernier phases, the X-ray photographs show many faint lines which vary with temperature. This means that the unit cell is complex and temperature sensitive. Recently, an ingenious solution to the structure has been found. The positions of the oxygen atoms are described by an incommensurate modulation wave. The modulation wave is such that the positions of the oxygen atoms represent a compromise between providing the pentagonal bipyramidal coordination demanded by the size of the large Ta atoms and the octahedral coordination needed to generate the correct Ta_2O_5 stoichiometry. When the oxygen atoms are distributed in accordance with the modulation wave, a percentage of the Ta atoms can reasonably be regarded as allocated to pentagonal bipyramidal coordination and others to octahedral coordination. Some, however, are not really in one or the other polyhedron. As the temperature increases, the modulation wave changes so as to increase the number of metal atoms in octahedral coordination at the expense of the pentagonal bipyramidally coordinated atoms. The modulation wave varies smoothly with temperature within the stability range of Ta_2O_5 and so the oxide has an *infinite number of structures available to it without any modification in its composition.*

Ta_2O_5 can form 'solid solutions' with a number of other oxides including Al_2O_3, TiO_2, WO_3 and ZrO_2. These are all incommensurate or modulated structures. As the composition varies, the modulation wavelength has to compromise between the composition (i.e. the number of oxygen atoms present) and the coordination preferences of the metal atoms to produce a stable structure. For example, if the Ta_2O_5 is reacted with WO_3 the modulation wave changes from that in the pure oxide so as to (i) accommodate more oxygen and (ii) accommodate more metal atoms (mainly W) in octahedral coordination. If ZrO_2 is the dopant, the wave takes into account that there is less oxygen in the structure and more atoms in pentagonal bipyramidal coordination. As before, the modulation wave also changes with temperature, producing the same remarkable state of affairs as in Ta_2O_5 itself, a single composition which can adopt an infinity of structures.

There are a number of non-stoichiometric compounds which show these incommensurate modulations quite clearly. However, the determination of the structures in detail is difficult and the problems posed by such materials at the forefront of structural crystallography.

9.13 Supplementary reading

Fortunately the subject matter of this chapter is well covered in a number of review articles. The classical article is that by A.D. Wadsley and should be consulted before the others. The textbook by Hyde and Andersson is next on the list, but all the articles cited contain different material and present different viewpoints of the subject.

A.D. Wadsley, *Non-stoichiometric Compounds*, ed. L. Mandelcorn, Academic Press, New York (1963).
B.G. Hyde and S. Andersson, *Inorganic Crystal Structures*, Wiley–Interscience, New York (1989).
J.S. Anderson, in *Solid State Chemistry*, N.B.S. Spec. Pub. 364, ed. R.S. Roth and S.J. Schneider, National Bureau Standards, Washington (1972), p. 295.
J.S. Anderson, *Defects and Transport in Oxides*, eds. M.S. Seltzer and R.I. Jaffee, Plenum Press, New York (1974).
J.S. Anderson, *J. Chem. Soc. Dalton Trans.* 1107 (1973).

B.G. Hyde, A.N. Bagshaw, S. Andersson and M. O'Keeffe, *Ann. Rev. Mater. Sci.* **4**, 43 (1974).
B.G. Hyde, S. Andersson, M. Bakker, C.M. Plug and M. O'Keeffe, *Prog. Solid State Chem.* **12**, 273 (1979).
E. Makovicky and B.G. Hyde, *Structure and Bonding* **46**, 101 (1981).
R.J.D. Tilley, in *The chemical physics of solids and their surfaces*, Vol. 8, eds. M.W. Roberts and J.M. Thomas, The Royal Society of Chemistry, London (1981).

The application of electron microscopy to non-stoichiometric compounds and a comparison of electron microscopy to other diffraction techniques for this purpose is covered in the following review articles in:

L. Eyring and A.K. Cheetham, *Nonstoichiometric Oxides*, ed. O.T. Sorensen, Academic Press, New York (1981).

Electron microscopy of minerals and good definitions of the terms used in mineralogy to describe non-stoichiometric compounds is given by:

D.R. Veblen, in *Reviews in Mineralogy*, chapter 6, Vol. 27, ed. P.R. Busek, Mineralogical Society of America (1992), p. 181.

Incommensurate structures were originally defined in terms of diffraction phenomena. A rather advanced review is given by:

T. Jansen and A. Janner, *Advances in Physics* **36**, 519 (1987).

Crystallographic aspects of incommensurate and modulated structures are stressed more in the textbook by Hyde and Andersson listed above and by:

S. van Smaalen, *Crystallogr. Rev.* **4**, 79 (1995).

A description of L–Ta_2O_5 as a modulated structure is given by:

S. Schmid, J.G. Thompson, A.D. Rae, B.D. Butler and R.L. Withers, *Acta Crystallogr.* **B51**, 698 (1995).
A.D. Rae, S. Schmid, J.G. Thompson and R.L. Withers, *Acta Crystallogr.* **B51**, 709 (1995).
S. Schmid, R.L. Withers and J.G. Thompson, *J. Solid State Chem.* **99**, 226 (1992).

10 Defects and non-stoichiometry in high temperature superconductors

10.1 Superconductivity and superconductors

Superconductivity was first discovered in 1911. It is a low temperature phenomenon which was first thought to be limited to a number of metallic elements and alloys. These materials are now known as *conventional superconductors*. The situation changed in 1987 when a number of complex copper oxides were synthesized which could transform to the super-conducting state above liquid nitrogen temperatures. As liquid nitrogen is in plentiful supply and relatively inexpensive, the possibility of much greater commercial application became a reality and laboratory experimentation became a relatively simple procedure. These materials are called *high temperature superconductors*.

When a superconducting material is cooled to such an extent that it enters the superconducting state it loses all electrical resistivity. The temperature at which this occurs is called the superconducting *transition temperature* and is written T_c. This means that in a material cooled to below T_c an electric current will flow in an unimpeded fashion forever, without decaying.

The superconducting state is also associated with notable magnetic behaviour. If a superconducting material is placed into a strong enough magnetic field it will revert to normal behaviour. The value of the field at this point is called the *critical magnetic field* strength, H_c. The way in which the normal state is restored can take place in two ways. In *Type I* superconductors the externally applied magnetic field is completely excluded from penetrating the superconductor (except for a thin surface layer) up to H_c. At this point, the superconducting properties are lost completely and the field penetrates the whole of the superconductor. When a Type I material is at room temperature any external magnetic field will pass through it in a normal way. However, when the material is cooled through the critical temperature the field is pushed out of the superconductor abruptly. The phenomenon of magnetic field exclusion is called the *Meisner effect*.

In Type II superconductors the magnetic field expulsion is less sharp. Above a critical field, H_{c_1}, the magnetic field starts to penetrate the superconductor. This results in cylinders of normal (non-superconducting) material embedded in the rest of the superconducting matrix. As the magnetic field increases the amount of normal material increases relative to the superconducting fraction until at the *upper critical field*, H_{c_2}, all the material becomes normal. The variation of the upper critical field with temperature is shown for some superconductors in Figure 10.1.

If a Type I or Type II superconductor is placed on a magnet, the magnetic field will be unable to enter the superconductor and the interaction can be strong enough to overcome gravitational forces and allow the superconductor to float above the surface of the magnet. This feature is known as *magnetic levitation*.

Although the potential applications for materials of this type are many, the usefulness of conventional superconductors was limited by the fact that the superconducting state could only be achieved quite close to absolute zero and at fairly low magnetic fields. The highest values for T_c with conventional superconductors remains at about 23 K to this day. Thus, the only applications for these materials are associated with equipment where the cost of cooling to such low temperatures, which requires expensive liquid helium, is offset by other benefits. Typically, such commercial applications consisted of magnetic resonance imaging equipment, of value in diagnostic

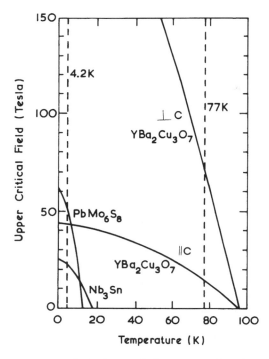

Figure 10.1 The temperature dependence of the upper critical field of a conventional superconductor, Nb₃Sn, the Chevrel phase PbMo₆S₈ and the high temperature superconductor YBa₂Cu₃O₇ both perpendicular and parallel to the crystallographic *c*-axis.

medicine, and superconducting quantum interference devices (SQUIDS), used for detecting minute changes in magnetic fields. Research applications include exploratory magnetically levitated trains and magnets in particle colliders. The lower cost of cooling high temperature superconductors makes them attractive for these applications.

One of the most interesting features of these new oxide superconductors is that they are all non-stoichiometric oxides. Moreover, the superconductivity is closely correlated with the degree of non-stoichiometry. This chapter will focus upon these vital aspects and apply much of the knowledge already presented to an understanding of this vital chemical side to these materials.

10.2 High temperature oxide superconductors

Although Cu-containing superconducting oxides have gained most attention, the phenomenon of superconductivity occurs in a number of oxide systems, as detailed in Table 10.1. These materials may provide important clues as to the chemical factors which lead to the superconducting state. All of these materials are *non-stoichiometric*. The presence or absence of superconducting behaviour is critically dependent upon the composition, as is T_c. This behaviour can be regarded as being due to the presence of chemical defects within the parent structure. This chapter will concentrate upon these chemical defects. Other defects in the materials, such as dislocations and grain boundaries, certainly do affect the current-carrying capabilities of the material but do not appear to control the appearance of superconductivity.

Table 10.1 Some superconducting oxides

Compound†	Structure	T_c (K)†	Active atom
$SrTiO_{3-x}$	Perovskite	0.5	Ti
$LiTi_2O_4$	Spinel	13	Ti
TiO	NaCl	0.6	Ti
NbO	NbO	1	Nb
K_xWO_3	HTB*	0.5	W
$BaPb_{0.75}Bi_{0.25}O_3$	Perovskite	13	Bi
$Ba_{0.6}K_{0.4}BiO_3$	Perovskite	31	Bi
$La_{1.85}Sr_{0.15}CuO_4$	K_2NiF_4	34	Cu
$YBa_2Cu_3O_{6.95}$	Perovskite layers	94	Cu
$Bi_2Sr_2CaCu_2O_8$	Perovskite layers	92	Cu
$Tl_2Ba_2Ca_2Cu_3O_{10}$	Perovskite layers	128	Cu
$HgBa_2Ca_2Cu_3O_8$	Perovskite layers	133	Cu

*Hexagonal tungsten bronze.
†Only representative formulae and T_c values are given.

The stoichiometric variation of greatest significance, with respect to superconductivity, is the oxygen content. A change in the oxygen content of a material must be accompanied by the presence of some sort of counter-defect to maintain charge neutrality. At present there is no theory that can account for the appearance of the superconducting state in these oxides and little guidance as to the true nature of the compensating defects.

10.3 Superconducting oxides which do not contain copper

10.3.1 $LiTi_2O_4$

$LiTi_2O_4$ was one of the first oxides to be found which showed super-conductivity, although the transition temperature is low, at about 13 K. $LiTi_2O_4$ adopts the normal spinel structure with Li^+ on the tetrahedral sites and Ti on the octahedral sites. In order to maintain charge balance, there must be equal numbers of Ti^{3+} and Ti^{4+} ions in the structure. This is often written in terms of an 'average valence', which for the Ti ions is 3.5. A *non-integral valence state* of this sort, we will see, is a key factor in all of the oxide superconductors. It appears to be necessary to endow one of the cations, the one which is presumed to be the cause of the superconductivity, with a partial valence.

As explained in chapter 6, the presence of two valence states readily leads to hopping conductivity. When electron transfer becomes very easy, the material would be metallic. This is the situation that exists in $LiTi_2O_4$. However, the spinel $LiTi_2O_4$ is able to form a solid solution in which additional Li substitutes for Ti on the octahedral sites, to yield a composition $Li_{1+x}Ti_{2-x}O_4$ with x taking values between 0 and 1/3. At the greatest extent of the solid solution, with a composition $Li_{4/3}Ti_{5/3}O_4$ all of the Ti ions are in the 4+ state and the material is an insulator. It seems that at intermediate compositions the material is composed of microdomains of superconducting $LiTi_2O_4$ coherently intergrown with non-superconducting $Li_{4/3}Ti_{5/3}O_4$. The transition temperature does not vary across the composi-tion range because the superconducting component never changes although the percentage of superconducting phase does alter.

10.3.2 $BaBiO_3$-related phases

The material $BaBiO_3$ has the *perovskite* structure. The charge on the Bi ions is not 4+ which is appropriate to the *perovskite* composition, but half of the atoms have a charge close to Bi^{3+} and half a charge close to Bi^{5+}. The result of this is that the oxygen octahedra surrounding the Bi atoms are of two sizes, with the Bi^{3+} in the large octahedra and Bi^{5+} in the smaller octahedra. By analogy with $LiTi_2O_4$, one would, therefore, expect this material to be metallic and perhaps a superconductor. However, because of the different geometries of the two types of octahedra, electron transfer is hindered and $BaBiO_3$ is a semiconductor. This is another general principle of use in materials chemistry, that is, *electron transfer will be hindered* if simultaneous *structural changes are also needed*. However, if these distortions can be suppressed it would be expected that the material might become metallic. This can be achieved via cation substitution and $BaBiO_3$ can actually be made into a superconductor. Moreover, $BaBiO_3$ is unique in that it can be made superconducting by cation substitution on either the Bi sub-lattice or the Ba sub-lattice.

The first superconductor prepared in this system was found in the $BaBiO_3$–$BaPbO_3$ system. The addition of $BaPbO_3$ allows a solid solution to form in which the Pb^{4+} ions substitute for Bi to form $BaPb_xBi_{1-x}O_3$. This solid solution remains semiconducting to a composition of about $BaPb_{0.6}Bi_{0.4}O_3$ at which composition it becomes metallic and remains so down to pure $BaPbO_3$. Superconductivity with a T_c of 13 K occurs in a narrow composition range close to $BaPb_{0.75}Bi_{0.25}O_3$.

The other $BaBiO_3$-derived superconductor is formed by substitution of Ba^{2+} by K^+ on the *perovskite A* sites. The narrow range of composition which superconducts is close to $Ba_{0.6}K_{0.4}BiO_3$ and shows a T_c of about 30 K.

All of these superconducting phases have one thing in common, mixed valence in one of the cations present. This appears to be a *vital requirement* and is a first clue in the hunt for the key to the occurrence of super-conductivity in oxides.

10.4   High-temperature superconducting copper oxides

Before embarking on a description of some of these materials, it is useful to note some of the features which apply to the group as a whole. Structurally, the phases are all related to the *perovskite* type structure illustrated earlier. The structure is redrawn somewhat differently in Figure 10.2. It shows that the structure can be thought of as being built up of a sequence of layers which we can label as AO and BO_2, where A are the large cations and B the medium-sized cations. In the copper oxide superconductors, the B atoms are Cu and the superconductivity resides in these CuO_2 sheets. However, note that in the structures described later not all copper–oxygen layers are identical and not all are superconducting sheets. The AO layers can be replaced by a variety of slabs, with structures mainly equivalent to slices of the *rock salt* and *fluorite* structures. These act as *charge reservoirs*, which are important in contributing to the conductivity process.

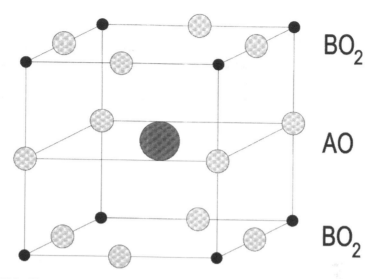

BO_2

AO

BO_2

Figure 10.2 The perovskite structure shown as a stacking of AO and BO_2 layers. (Compare with Figure 6.5.)

A second important factor in these materials is that the copper valence (in most compounds) must lie between the formal values of Cu^{2+} and Cu^{3+}. It seems to be a fairly common observation that the maximum value of the transition temperature, T_c, occurs when the average charge has a value of about 2.33. The aim of many experiments is to modify the AO regions to achieve this.†

10.5 La$_2$CuO$_4$ and related phases

10.5.1 La$_2$CuO$_4$

The phase La$_2$CuO$_4$ contains trivalent La and divalent Cu, and adopts a slightly distorted version of the K_2NiF_4 structure. This results in the compound being orthorhombic at room temperature with the a-axis slightly shorter than the b-axis. At temperatures above about 260 °C, the structure becomes tetragonal with the a- and b-axes equal. The La$_2$CuO$_4$ structure contains sheets of the *perovskite* type, one CuO_6 octahedron in thickness stacked up one on top of the other, as can be seen in Figure 10.3. It is also

---

†There is considerable uncertainty as to whether the defects should be Cu^{3+} or O^-, that is to say, whether the holes should sit on copper cations or anions. In this chapter the Cu^{3+} convention will be chosen. (See also the first footnote in Chapter 6.)

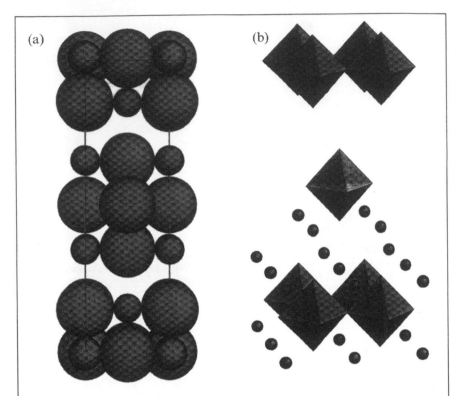

Figure 10.3   The K_2NiF_4 structure. (a) A unit cell showing F^- ions as large spheres, Ni^{2+} ions as medium-sized spheres and K^+ ions as smallest spheres. (b) The structure shown in terms of NiF_6 octahedra and in the lower part of the diagram NiF_6 octahedra and sheets of composition KF.

The K_2NiF_4 structure

The K_2NiF_4 structure is made up of sheets of the *perovskite* type-one octahedra in thickness, stacked up one on top of the other, as can be seen in Figure 10.3. The unit cell is tetragonal with $a = 0.4006$ nm, $c = 1.3706$ nm. There are two formula units in a unit cell. The positions of the K and F atoms in the KF layers is similar to the cation and anion positions in the *rock salt* structure and so the structure is often said to made up of an *intergrowth* of *perovskite* and *rock salt* slabs.

A number of oxides crystallize with this structure including La_2NiO_4, Sr_2TiO_4 and Sr_2SnO_4.

easily described in terms of stacked CuO_2 and LaO layers, as the figure suggests.

When prepared in air by heating oxides, the compound is usually stoichiometric with the oxygen content close to 4.00. However, small changes in the oxygen content can be easily introduced by heating in oxygen at higher or lower pressures, to make materials with a formula $La_2CuO_{4+\delta}$. For values of δ which are negative, indicating an oxygen loss, the material is a poor semiconductor. Notionally, the defect structure consists of oxygen vacancies with charge compensation being by way of two Cu^+ ions per oxygen vacancy, but this has not been confirmed experimentally. For values of δ which are positive, indicating an excess of oxygen, the electronic properties change drastically. First, the material becomes a metal and then a superconductor. The maximum value of T_c observed is 38 K in $La_2CuO_{4.13}$.

The structure of the oxygen-rich material is quite complex. At the temperatures at which these compounds are prepared, the extra oxygen atoms are introduced into the structure as interstitials which occupy sites midway between the LaO planes. They are also arranged so as to be as far away from the La^{3+} ions as possible, and sit more or less at the centre of an La^{3+} tetrahedron. For each interstitial introduced, two Cu^{2+} ions become Cu^{3+} ions. This is the same as saying that two holes are created per oxygen interstitial and that these are located on Cu^{2+} ions, at least in the semiconducting state. It is these holes that are the charge carriers in the superconducting state and the material is termed a 'hole' superconductor. At room temperature, the oxygen interstitials are not distributed at random. Instead they segregate into only some of the LaO layers. These defect-containing layers can stack up in several different ways to give several different crystallographic repeat lengths along the c-axis of the basic La_2CuO_4 unit cell. The actual configurations observed vary with oxygen content and with temperature. On cooling a typical preparation, the microstructure of the sample consists of microdomains of material with no interstitial oxygen atoms, together with material with several different ordering patterns of these interstitial oxygen atom layers to give quite a complex and interesting non-stoichiometric structure.

10.5.2 Substituted La_2CuO_4 phases

The generation of Cu^{3+} is not especially easy in La_2CuO_4 and it is much easier to utilize the technique of valence induction. In fact, it was this approach which lead to the discovery of the first copper oxide super-conductor, $La_{2-x}Ba_xCuO_4$. Since then, it has been shown that the incorporation of Sr, Ca, Na and K also induce superconductivity in this material. The material which has been investigated most has the formula $La_{2-x}Sr_xCuO_4$, which shows a maximum transition temperature of 37 K at a composition of $La_{1.85}Sr_{0.15}CuO_{4-\delta}$. In this compound, the impurity Sr^{2+} cations substitute for La^{3+}. Because the Sr^{2+} ion has a lower charge than the La^{3+}, some mechanism for maintaining charge neutrality must be found. This can take place in one of two ways. It is possible to generate one oxygen vacancy for every two Sr^{2+} added, to give a formula $La_{2-x}Sr_xCu^{2+}O_{4-x/2}$. Alternatively, one Cu^{3+} could form for each Sr^{2+} substituted, to give a formula $La_{2-x}Sr_xCu^{2+}_{1-x}Cu^{3+}_xO_4$.

The balance between these two alternative defects is very delicate and leads to a most surprising situation. Initially, the Cu^{3+} option is preferred. Because Cu^{3+} can be looked on as Cu^{2+} together with a trapped hole, it is considered that a hole population is present which leads to a super-conducting state. As the Sr^{2+} concentration rises so does the Cu^{3+} concentration, up to a peak when x is approximately 0.2. As more Sr^{2+} is added, the preferred defect now becomes the oxygen vacancy and the oxygen content of the parent phase falls below 4.0. The number of Cu^{3+} ions decreases as oxygen vacancies form, and when the concentration of Sr^{2+} reaches approximately 0.32 all of the compensation is via vacancies and the material is no longer a superconductor! This is shown in Figure 10.4.

The same effect is also found if the $La_{1.85}Sr_{0.15}CuO_4$ is heated at about 500 °C in a vacuum. This causes oxygen loss and a consequent loss of superconductivity. Reheating in oxygen restores the superconducting prop-

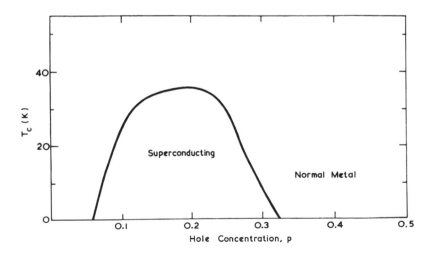

Figure 10.4   A diagram of the dependence of T_c on the Sr^{2+} content in $La_{2-x}Sr_xCuO_4$. Initially Sr^{2+} substitution introduces Cu^{3+} (i.e. holes) into the structure. At higher concentrations, the Cu^{3+} population is replaced by oxygen vacancies and the material loses its superconducting properties.

erties. Thus, it is seen that the nature of the defects in the phase completely controls the superconductivity.

10.5.3 $La_{n+1}Cu_nO_{3n+1}$ and related phases

La_2CuO_4 can be considered to be the first member of a homologous series of layered perovskite oxides described by the formula $La_{n+1}Cu_nO_{3n+1}$. These oxides contain thicker *perovskite* slabs than La_2CuO_4, and the value of n in the formula gives the number of octahedra in the *perovskite*-like slabs as shown in Figure 10.5. They can also be regarded as ordered intergrowths between La_2CuO_4 and the *perovskite* $LaCuO_3$. This latter phase contains only Cu^{3+} rather than the more stable Cu^{2+} ion and so the homologous series can only be formed at high oxygen pressures.

Despite this limitation, a closely related group of compounds can form in air. These can be typified by $La_2SrCu_2O_6$. The structure of this phase is shown in Figure 10.6. It is seen to be very similar to the structure of the $n = 2$ phase $La_3Cu_2O_7$. However, the Cu is in the divalent state, achieved by substitution of one La^{3+} by Sr^{2+} and by removal of a layer of oxygen to convert the Cu coordination from octahedra to square pyramidal. This oxide and similar phases containing other alkaline earth metals are remarkable because they can take up and give out oxygen very easily. $La_2SrCu_2O_6$ can easily and reversibly reach the composition of La_2Sr-$Cu_2O_{6.2}$. This introduces Cu^{3+} into the compound and one would expect that these would become hole superconductors. This is not so for either $La_2SrCu_2O_{6+\delta}$, or $La_2CaCu_2O_{6-\delta}$, but the mixed phase

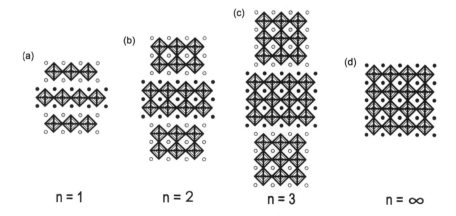

Figure 10.5 The idealized structures of the $La_{n-1}Cu_nO_{3n+1}$ phases. The shaded squares represent CuO_6 corner-linked octahedra and the circles represent the La atoms. (a) The idealized K_2NiF_4 or La_2CuO_4 structure and (d), in which $n = \infty$, corresponds to the ideal *perovskite* structure, $LaCuO_3$.

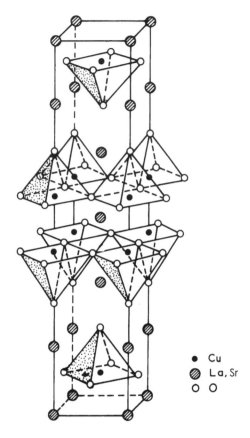

Figure 10.6   The idealized structure of the oxide $La_2SrCu_2O_6$. The small, filled circles represent Cu atoms which reside at the centres of CuO_5 square pyramids. The large, open circles represent O atoms and the large, shaded circles represent the La or Sr atoms, which cannot be differentiated.

$La_{1.6}Sr_{0.4}CaCu_2O_{6+\delta}$ can be made to superconduct with a T_c of about 60 K. This reveals the complexity of the superconducting process and the difficulties associated with the prediction of which materials might show this property.

10.6 Nd$_2$CuO$_4$ electron superconductors

The structure of Nd_2CuO_4 is shown in Figure 10.7. A comparison of Figures 10.3 and 10.7 reveals that the principle difference between the Nd_2CuO_4 and La_2CuO_4 structures lies in the disposition of the oxygen atoms as the cations in the two structures are in almost identical positions. However, in Nd_2CuO_4 the Nd^{3+} ions are in the centres of oxygen cubes and so this

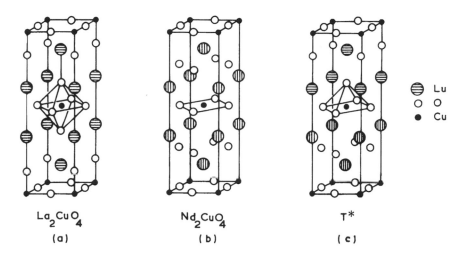

Figure 10.7 (a) The La_2CuO_4 structure. (b) The Nd_2CuO_4 structure. (c) The hybrid T^* structure which is an intergrowth of half unit cells of the La_2CuO_4 and Nd_2CuO_4 types.

region of the structure can be likened to slabs of the *fluorite* type and the structure is often described as an *intergrowth* of *perovskite* and *fluorite* structures. The Cu atoms lie at the centres of square coordination groups and between these are *fluorite* slabs.

Like the lanthanum copper oxides, this compound can lose oxygen down to at least a composition of $Nd_2CuO_{3.5}$. The oxygen deficit is due to oxygen vacancies and these are balanced by the formation Cu^+ ions in the structure. These reduced materials are insulators.

However, Nd_2CuO_4 can be made to superconduct by a substitution of the lanthanide by a *higher valence cation* such as Ce^{4+} to form $Nd_{2-x}Ce_xCuO_{4-\delta}$. If we use the same terminology as before, it is possible to write the formula as $Nd_{1-x}Ce_xCu^{2+}_{1-x}Cu^+_xO_{4-\delta}$ and to suppose that one Cu^+ ion is created for every Ce^{4+} in the crystal. The Cu^+ ion is equivalent to a Cu^{2+} ion together with a trapped electron and so the structure is regarded as possessing an *excess of electrons* in the structure. The compounds turn out to be *electron superconductors* rather than hole superconductors as in the case of the La_2CuO_4-related phases. In a formal way, the superconducting state occurs over a rather narrow range of substitutions with x taking values of approximately 0.12–0.18. The maximal T_c values found for these phases is 24 K, reached in the compound $Nd_{1.85}Ce_{0.15}CuO_4$.

It is noteworthy that hole superconduction does not seem possible in this phase, as Nd^{3+} substitution by Ca^{2+}, Sr^{2+} or Ba^{2+} does not give superconducting phases.

10.7 $YBa_2Cu_3O_7$ and related phases

The compound $YBa_2Cu_3O_7$, sometimes referred to as '123', which describes the cation ratios of 1Y:2Ba:3Cu, has been widely studied because it is relatively easy to prepare and it was the first superconductor discovered with a T_c above the boiling point of liquid nitrogen. The crystal structure of $YBa_2Cu_3O_7$, shown in Figure 10.8, is closely related to that of the *perovskite* type. The unit cell consists of three *perovskite*-like cells stacked one on top of the other, as can be seen from the cation stacking shown in Figure 10.8(a). The middle *perovskite* unit contains Y as the large A cation and Cu as the smaller B cation. The cells above and below this contain Ba as the A cation and Cu as the B cation, to give a metal formula of YBa_2Cu_3 as one would expect for a tripled *perovskite* cell, $A_3B_3O_9$. The unit cell of the superconductor should contain nine O atoms. Instead, the seven O atoms present are arranged in such a way as to give the Cu atoms square pyramidal coordination, shown in Figure 10.8(b) and square planar coordination, shown in Figure 10.8(c), rather than entirely octahedral as in the normal *perovskites*. This results in an orthorhombic structure at room temperature. If the ions are allocated the normal formal charges of Y^{3+}, Ba^{2+} and O^{2-}, the Cu must take an average charge of 2.33, which can be considered to arise from the presence of two Cu^{2+} and one Cu^{3+}.

The appearance of superconductivity in this material is closely related to the oxygen content, which, like La_2CuO_4, is a hole superconductor. In fact, for the exact composition $YBa_2Cu_3O_{7.0}$ the material does not show superconducting behaviour. This only appears when a small amount of oxygen is lost. In reality, the compound can readily lose oxygen down to a composition of $YBa_2Cu_3O_{6.0}$ and the way in which the superconductivity changes over this composition range has been well documented and is shown in Figure 10.9. The maximum value of T_c, close to 94 K, is found at the composition of $YBa_2Cu_3O_{6.95}$. As more oxygen is removed, the value of T_c falls to a plateau of approximately 60 K when the composition lies between the approximate limits of $YBa_2Cu_3O_{6.7}$ and $YBa_2Cu_3O_{6.5}$. Continued oxygen removal down to the phase limit of $YBa_2Cu_3O_{6.0}$ rapidly leads to a loss of superconductivity. However, it is necessary to introduce a note of caution, because the exact behaviour of any sample depends upon the defects present and this in turn can depend upon whether the samples are cooled quickly or slowly.

The oxygen atoms are not lost at random in this reduction but come from the CuO_4 square planar units to convert the copper coordination to linear, as can be seen by comparison of Figures 10.8(a) and (b) with 10.8(c). This suggests that oxidation in *particular regions* of the unit cell is important. If any of these intermediate compositions are carefully heated, the oxygen vacancies order into a number of superstructures and at fixed compositions. The principle oxygen compositions which have been reported as producing

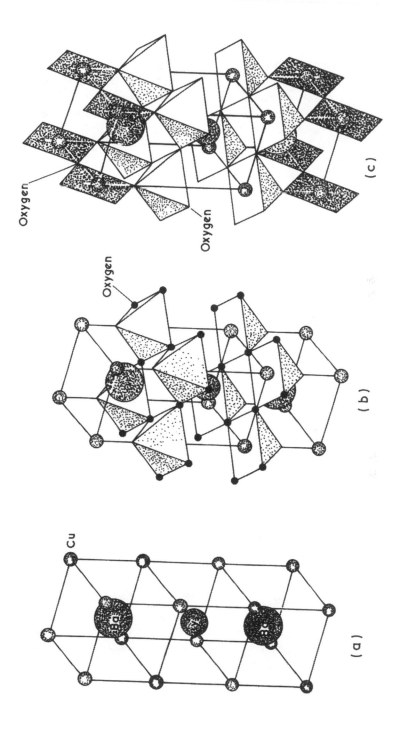

Figure 10.8 (a) The metal atom positions in $YBa_2Cu_3O_7$ and $YBa_2Cu_3O_6$, which are identical to those found in three unit cells of the *perovskite* type. (b) The idealized structure of $YBa_2Cu_3O_6$ and (c) $YBa_2Cu_3O_7$.

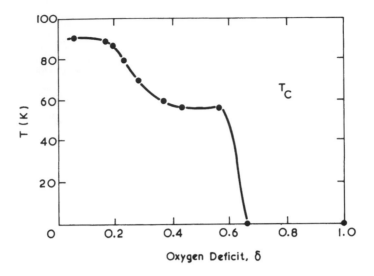

Figure 10.9 The superconducting transition temperature, T_c, as a function of the oxygen deficit δ for $YBa_2Cu_3O_{7-\delta}$.

long-range ordered microstructures are $YBa_2Cu_3O_{6.875}$, $YBa_2Cu_3O_{6.50}$ and $YBa_2Cu_3O_{6.125}$. At intermediate compositions, microdomains of these ordered structures can intergrow easily with one another so as to accommodate the non-stoichiometry in an ordered fashion. The effect of these ordered arrangements on T_c remains to be fully explored.

Besides the change in the superconducting properties, oxygen loss also changes the symmetry of the structure. The introduction of the oxygen vacancies tends to make the cell become tetragonal. Oxygen is easily lost by heating the samples to temperatures of several hundred degrees. Thus, the tetragonal form of this compound has an extended composition range which is dependent on oxygen content and temperature. Although the range of the orthorhombic and tetragonal structures has been carefully mapped, the oxygen content remains the most important factor in endowing the material with superconducting properties.

These tetragonal and orthorhombic forms do have an important structural effect. $YBa_2Cu_3O_{7.0}$ is normally prepared at temperatures of up to 950 °C. At these temperatures the material is tetragonal, with the a-axis equal to the b-axis. On cooling, either of these axes can become the a- or b-axis of the orthorhombic form. This produces random twinning on (110) planes and most crystals at room temperature are heavily twinned. There remains some uncertainty about the effect of these twins on the magnitude of T_c. However, they play an important role in decreasing the current-carrying capability of these compounds and they make device fabrication difficult.

It is clear that the oxygen content of the material is critical. Thus, control of the oxygen content is vital during device fabrication. One of the most important parameters, in this respect, is the oxygen diffusion coefficient. Examination of Figure 10.8(c) will suggest that the diffusion coefficients along the three axes will be different. Diffusion turns out to be fastest along the b-axis, and very much slower along the c-axis, with the a-axis diffusion coefficient rather smaller than the b-axis value. Because of this marked anisotropy, it is possible to measure tracer diffusion coefficients along the b-axis of a plate without worrying too much about diffusion along the other axes. The tracer used is usually O^{18} and the experiments reported to date have settled close to the values

$$D^* = 1.3 \times 10^{-3} \exp\left(-\frac{119}{RT}\right)$$

where the pre-exponential factor is in $m^2 s^{-1}$ and the activation energy is in $kJ \, mole^{-1}$. Some experimental results for O^{18} tracer diffusion are shown in Figure 10.10(a).

Although the tracer diffusion coefficient is of interest, in practical work the chemical diffusion coefficient is of more importance. We have seen that the relationship between the chemical diffusion coefficient $\tilde{D}$ and the tracer diffusion coefficient D^* is given by an expression of the form

$$\tilde{D} = D^* F$$

where F is the thermodynamic coefficient. It has proved possible to calculate the thermodynamic term from the oxygen pressure over the sample and the oxygen content of the solid. The result, given in Figure 10.10(b), shows that the thermodynamic coefficient increases rapidly close to an oxygen content of 7.0. Thus, although the tracer diffusion remains constant, surprisingly it seems that the chemical diffusion of oxygen will increase as the fully oxidized composition is approached. Some values of the chemical diffusion coefficient are shown in Figure 10.10(c).

The complexity of this non-stoichiometric oxide is considerably increased when cation substitutions are considered. In general, most of the lanthanides can replace Y and superconducting non-stoichiometric phases result. The only exception is $PrBa_2Cu_3O_7$, which has not been made superconducting to date. The substitution of Cu by other metal ions, such as Ni and Zn, is also possible. These substitutions invariably decrease the value of T_c. This type of substitution and also allow the oxygen content to increase above 7.0. This happens, for example, when some Cu is replaced by Co, when a composition as high as $YBa_2Co_{0.8}Cu_{2.2}O_{7.4}$ can be achieved.

These substitutions suggest that Cu alone gives the highest value of T_c in a material.

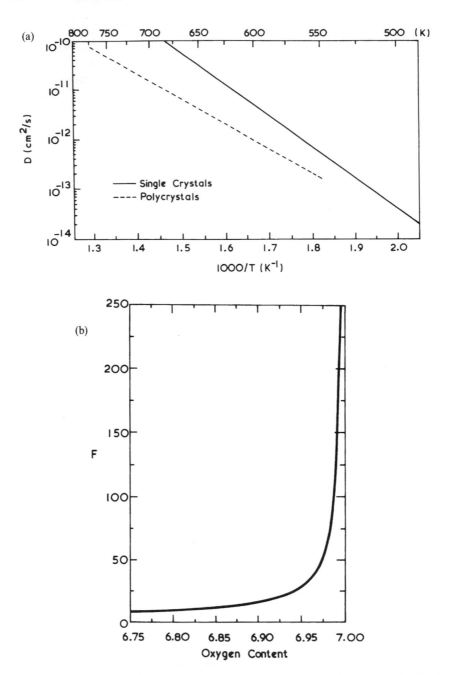

Figure 10.10   Oxygen difffusion data for orthorhombic $YBa_2Cu_3O_7$. (a) An Arhenius plot for O^{18} tracer diffusion. (b) Variation of the thermodynamic factor F as a function of composition. (c) Some values of the chemical diffusion coefficient. [Redrawn from data given by Conder, Kruger and Kaldis, *Perspectives in Solid State Chemistry*, ed. K.J. Rao, Narosa, New Delhi (1995).]

(c)

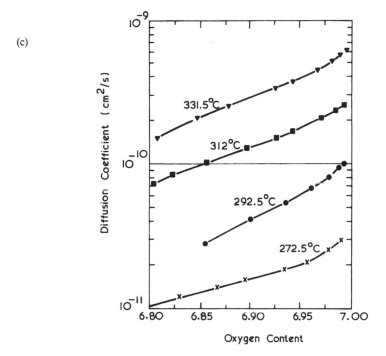

Figure 10.10   *Continued.*

10.8 $Pb_2Sr_2YCu_3O_8$

In the previous sections, the key route to producing a superconducting material tended to be oxidation. The reason for including $Pb_2Sr_2YCu_3O_8$ in this chapter is to highlight the fact that this oxidation has to be *directed at quite specific regions* within the crystal structure. The compound $Pb_2Sr_2YCu_3O_8$ has a structure shown in Figure 10.11. The familiar architecture of CuO_2 planes, as part of *perovskite*-like regions of structure, is apparent. The formal charges on the atoms are Pb^{2+}, Sr^{2+}, Y^{3+} and O^{2-}, resulting in an average Cu charge of $+5/3$, probably distributed as two Cu^{2+} and one Cu^+ ion. Extrapolation from the behaviour of other phases suggests that oxidation would generate Cu^{3+} and superconducting behaviour would result. This appears to pose no structural problems as there is plenty of space for oxygen incorporation. However, a super-conducting transition does not occur even though oxygen uptake to a composition of about $Pb_2Sr_2YCu_3O_{9.5}$ is easily possible. What in fact happens, is that the oxygen is incorporated into the Cu planes, rather than the CuO_2 planes as anticipated, and the additional positive charges needed to maintain charge neutrality are localized on the Pb^{2+} ions to form Pb^{4+}. The extra oxygen changes the coordination polyhedra around the bulky

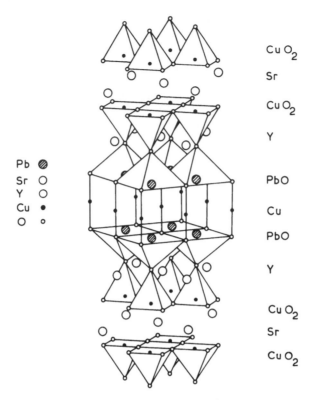

Figure 10.11   The $Pb_2Sr_2YCu_3O_8$ structure showing the Cu^{2+} ions in square pyramidal oxygen co-ordination and the Cu^+ linearly co-ordinated by oxygen. The sequence of planes in the structure is listed to the right of the figure.

Pb^{2+} ions which transform to much smaller Pb^{4+} ions and as a result the additional charges are quite localized. This is similar to the situation holding in $BaBiO_3$ where charge transfer was inhibited because of the different coordination polyhedra around the two different Bi ions.

The problem is how to induce *extra charges* in the CuO_2 layers. This has been achieved by replacing some of the Y^{3+} by Ca^{2+}. Figure 10.11 shows that the Y^{3+} ions reside between the important CuO_2 layers. Replacement of some Y^{3+} by Ca^{2+} results in some of the Cu^{2+} ions in the adjacent CuO_2 layers transforming to Cu^{3+} ions to form a superconducting material. The highest value of T_c has been found when half of the Y has been replaced, to give a composition $Pb_2Sr_2Y_{0.5}Ca_{0.5}Cu_3O_8$. If this superconducting compound is heated in oxygen, the oxygen content increases and the compound reverts to a non-superconducting state.

This material demonstrates an important aspect of this detailed crystal engineering. Not only must we induce a mixed valence for the Cu ions, but it must be between Cu^{2+} and Cu^{3+} and these must be localized in the same layer or adjacent layers to lead to the onset of superconducting behaviour. It also reveals that one important aspect in making Cu such a useful tool is that this ion can take several valence states in a number of different coordination geometries without any difficulties.

10.9 The Bi, Tl and Hg homologous series of superconductors

As soon as the high T_c compound $(La,Ba)_2CuO_4$, described earlier, was found, other systems which were likely to produce similar materials were investigated. A very likely candidate was the compound Bi_2CuO_4. Although the structure of this phase is different from that of La_2CuO_4 it seemed reasonable to try to replace some of the Bi^{3+} by a divalent cation in order to induce the formation of Cu^{3+} in the material and hopefully produce a new superconductor. These experiments were not successful, but nevertheless they did lead to the discovery of a series of superconducting phases which illustrate the marked dependence of T_c on the thickness of the *perovskite*-like regions in the compounds. The same feature is shown in a structurally similar series of oxides in which Tl or Hg replace Bi. These are listed in Table 10.2.

The first superconducting oxide to be made in this system was $Bi_2Sr_2CuO_6$. It was subsequently shown that this compound was the $n = 1$ member of a new homologous series of phases of general formula $Bi_2 Ca_{n-1}Sr_2Cu_nO_{2n+4}$, with n taking values from 1 to 3. Because of the rather complex formulae, these phases are often referred to in a shorthand notation which specifies the cation ratios in the order BiCaSrCu. The $n = 1$ phase is 2021, i.e. made up from **2Bi:0Ca:2Sr:1Cu**. The $n = 2$ phase is written 2122 and the $n = 3$ phase is 2223, and so on. These labels are given for all phases in Table 10.2.

Table 10.2   The Bi, Tl and Hg homologous series of superconductors

n	Formula	Notation	T_c (K)	n	Formula	Notation	T_c (K)
Double layers							
1	$Tl_2Ba_2CuO_6$	2021	92	1	$Bi_2Sr_2CuO_6$	2021	10
2	$Tl_2CaBa_2Cu_2O_8$	2122	119	2	$Bi_2CaSr_2Cu_2O_8$	2122	92
3	$Tl_2Ca_2Ba_2Cu_3O_{10}$	2223	128	3	$Bi_2Ca_2Sr_2Cu_3O_{10}$	2223	110
4	$Tl_2Ca_3Ba_2Cu_4O_{12}$	2324	119				
Single layers							
1	$TlBa_2CuO_5$	1021		1	$HgBa_2CuO_4$	1021	94
2	$TlCaBa_2Cu_2O_7$	1122	103	2	$HgCaBa_2Cu_2O_6$	1122	127
3	$TlCa_2Ba_2Cu_3O_9$	1223	110	3	$HgCa_2Ba_2Cu_3O_8$	1223	133
				4	$HgCa_3Ba_2Cu_4O_{10}$	1324	126

The structure of the $n = 1$ compound, $Bi_2Sr_2CuO_6$, shown in Figure 10.12, is very similar to La_2CuO_4, with a layer of corner-shared CuO_6 octahedra with a single *perovskite* layer being a prominent feature. These are connected by a layer of composition Bi_2O_2 which is shown as a planar sheet in Figure 10.12. The other members of this homologous series have thicker

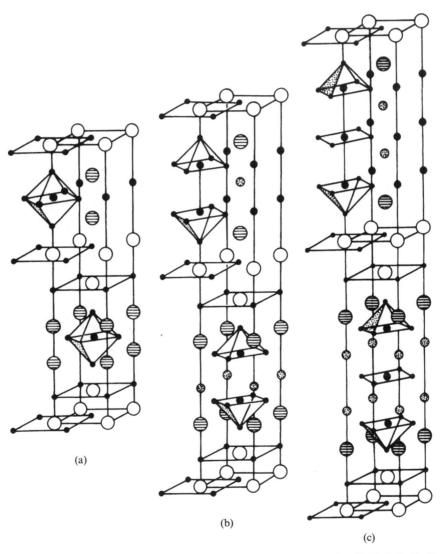

(a)

(b)

(c)

Figure 10.12   The idealized structures of (a) $Bi_2Sr_2CuO_6$ and $Tl_2Ba_2CuO_6$; (b) $Bi_2CaSr_2Cu_2O_8$ and $Tl_2CaBa_2Cu_2O_8$; and (c) $Bi_2Ca_2Sr_2Cu_3O_{10}$ and $Tl_2Ca_2Ba_2Cu_3O_{10}$. The Cu atoms are in octahedral (a) or square pyramidal oxygen coordination (b) and (c). Oxygen ions are represented by the smallest filled circles, copper ions by medium filled circles, bismuth ions by open circles and alkali metals by horizontal and dot shaded circles

perovskite layers, two in the case of $Bi_2CaSr_2Cu_2O_8$ and three in the case of the $Bi_2Ca_2Sr_2Cu_3O_{10}$, but all have the same Bi_2O_2 separating slabs, as shown in Figure 10.12(b) and (c). The perovskite layers need a large cation both to balance the charges present and to prevent the structure from collapsing and in these oxides this role is taken up by Ca, which is the interpolated atom between the CuO_2 sheets. In addition, the CuO_6 octahedra are transformed into CuO_5 square pyramids in the *perovskite* layers of the $n = 2$ and $n = 3$ phases. These structures are shown in Figure 10.12. An almost identical series of phases forms with Tl instead of Bi.

There are a number of unique points to mention concerning these phases. All need a slight oxygen excess to show superconductivity and the formulae are more accurately written in the form $Bi_2Ca_{n-1}Sr_2Cu_nO_{2n+4+\delta}$. When these compounds are prepared in air this extra oxygen is incorporated directly and the process is sometimes referred to as *self doping*. If the samples are treated so as to remove this extra oxygen, then a loss of superconductivity results.

The most striking and important aspect of these compounds is the relationship between the thickness of the *perovskite* sheets, given by n in the series formulae, and the superconducting transition temperature, T_c. These are set out in Table 10.2. It is seen that T_c increases as the value of n increases, giving rise to hopes that even higher T_c values could be achieved simply by increasing n beyond 3. Unfortunately, compounds with n much higher than 3 are not easy to synthesize. In fact, even the compound $Bi_2Ca_2Sr_2Cu_3O_{10}$ is extremely difficult to make, but can be 'stabilized' by replacing some of the Bi by Pb to form compounds with compositions typified by $Bi_{1.6}Pb_{0.4}Ca_2Sr_2Cu_3O_{10}$. However, several materials corresponding to $n = 4$, especially $Tl_2Ca_3Ba_2Cu_4O_{12}$, have been made and these do not have the hoped for higher T_c values. It seems that in the compounds made to date, T_c peaks with the $n = 3$ phases.

Bonding between the Bi_2O_2 layers and the *perovskite*-like slabs is weak, which results in the crystals easily flaking, rather like mica. This weak bonding also allows the geometry of the Bi_2O_2 sheets to be relatively unconstrained by the geometry of the adjacent *perovskite*-like slabs. As noted, the oxygen content is not identical to that given by the idealized series formulae and in these phases the oxygen excess is accommodated within the Bi_2O_2 regions. This alters the dimensions of these units slightly and results in an *incommensurate superlattice* along the *b*-direction, because the repeat spacing of the Bi_2O_2 units no longer exactly matches that of the *perovskite* layers. This structural complexity does not affect the superconducting properties, which provides strong evidence for the assumption that the superconductivity is closely associated with the CuO_2 sheets buried within the *perovskite* slabs.

Although the structures of the Tl and Bi phases are similar, there are some aspects which differ. The smaller size of the Tl^{3+} ion necessitates the replacement of Sr^{2+} by the larger Ba^{2+} in order to maintain structural stability. The homologous series formula is, therefore, $Tl_2Ca_{n-1}Ba_2Cu_nO_{2n+4}$. In addition the bonding between the *perovskite* layers and the Tl_2O_2 sheets is stronger than in the Bi case. This means that the oxides do not cleave so readily and the incommensurate superlattice does not easily occur. It also means that the oxygen deficit is not quite so easily accommodated within the Tl_2O_2 layers. A consequence of this is that another method of taking in a change of composition is helpful. This seems to be by way of intergrowths of the various members of the series, and such disordered and ordered intergrowths are quite a common feature of the Tl-containing crystals.

A more significant difference is the existence of a second homologous series of oxides which does not form in the Bi series. In these, the *perovskite* slabs are joined together by single TlO sheets rather than double Tl_2O_2 slabs. The homologous series formula is $TlCa_{n-1}Ba_2Cu_nO_{2n+3}$ and the structures of these compounds are shown in Figure 10.13. Apart from this structural change, they behave in quite a similar way to the double-layer phases. In particular, a small oxygen excess is needed to induce super-conducting behaviour.

The most recent series of high temperature phases to be found have a formula $HgCa_{n-1}Ba_2Cu_nO_{2n+2}$ and structures which are very similar to the TlO phases just described. The structures are derived from those of the Tl-containing phases by replacing the TlO layers in these latter compounds by sheets of Hg atoms, as shown in Figure 10.13. As in the case of the other members of this group, the value of T_c varies with the value of n in the series formula, up to a maximum at the $n = 3$ compound $HgCa_2Ba_2Cu_3O_8$ of 133 K. The major interest in these phases is that, to date, this is the highest T_c value yet reported for materials at normal pressures. The $n = 4$ member, $HgCa_3Ba_2Cu_4O_{10}$, has a T_c value of 126 K. As in the Bi and Tl compounds, an oxygen excess over the nominal formula is required to induce the onset of superconductivity. In these phases, this is incorporated into the Hg planes to convert them to HgO_δ planes.

10.10 Conclusions

Not all the known high temperature superconductors have been mentioned in this chapter. However, the examples given provide the basis for an understanding of the factors which must be controlled in order to make these materials, at least in the laboratory. These appear to be

1. The presence of mixed cation valence states.
2. A range of oxygen non-stoichiometry.

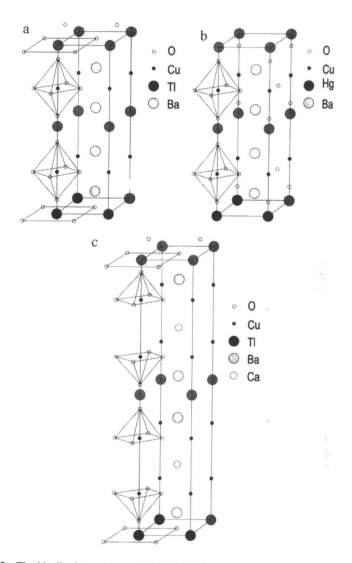

Figure 10.13   The idealized structures of (a) TlBa$_2$CuO$_5$; (b) HgBa$_2$CuO$_4$ (c) TlCaBa$_2$Cu$_2$O$_7$. The Cu atoms are in octahedral (a) and (b) or square pyramidal oxygen coordination (c).

3.   The ability of the 'superconducting' cation to adopt to a variety of different coordination polyhedra.

One important aspect in making Cu such a useful cation, in this respect, is that this ion can take several valence states in a number of different coordination geometries without any difficulties.

Although these guidelines are helpful, there are still large numbers of interesting and exciting problems remaining. For example, it is not known why the Hg superconductors have a higher T_c than any others, and there is still no guide as to whether materials with even higher T_c values can be fabricated. In addition, there is still no theoretical basis for understanding these phases. The story of these remarkable materials and the defect chemistry associated with them still has a long way to go.

10.11 Supplementary reading

Since the discovery of high-temperature superconductivity there has been a deluge of papers, numbering tens of thousands in just a few years. The list below gives just a few review articles and sources of wider interest.

The following articles give a good overview of the subject at a readable level:

The discovery of (conventional) superconductivity, *Sci. Am.* March, 84 (1997).
Perovskites in relation to non-stoichiometry and high-temperature superconductivity, *Sci. Am.* June, 52 (1989).
Applications of high temperature superconductors, *Sci. Am.* February, 45 (1989).
Crystal chemical aspects of high temperature superconductors, *Sci. Am.* August, 24 (1990).
SQUIDs, superconducting quantum interference devices, *Sci. Am.* August, 36 (1994).

The physics of superconductivity is covered in:

J.R. Waldron, *Superconductivity of Metals and Cuprates*, Institute of Physics, Bristol (1996).

The relationship between electronic conductivity, chemical bonding and structure in oxides, including oxide superconductors, is given by:

P.A. Cox, *Transition Metal Oxides*, Oxford University Press, Oxford (1992).
P.A. Cox, *The Electronic Structure and Chemistry of Solids*, Oxford University Press, Oxford (1987).

An up-to-date tabulation of superconductors, both conventional and high temperature, will be found in the current edition of the *Handbook of Chemistry and Physics*, CRC Press, Boca Raton, FL, updated approximately annually.

Some review articles, which also give a flavour of the rapid progress made in the science and engineering of these remarkable compounds are:

MRS Bull. **XIV**, January (1989).
B. Raveau, C. Michel and M. Hervieu, *J. Solid State Chem.* **88**, 140 (1990).
MRS Bull. **XV**, June (1990).
M. Marezio, *Acta Crystallogr.* **A47**, 640 (1991).
MRS Bull. **XVII**, August (1992).
P.L. Gai and J.M. Thomas, *Superconductivity Rev.* **1**, 1 (1992).
MRS Bull. **XIX**, September (1994).
C.N.R. Rao and A.K. Ganguly, *Acta Crystallogr.* **B51**, 604 (1995).

The diffusion data shown in Figure 10.10 were redrawn from information in:

K. Condor, C. Kruger and E. Kaldis, *Perspectives in Solid State Chemistry*, ed. K.J. Rao, Narosa, New Dehli (1995).

11 Non-stoichiometry: an overview

11.1 Ordering, assimilation and elimination of defects

Throughout this book the complexity of the defect structures encountered has increased. It is useful to integrate these changes into a coherent picture. At the outset, the modes of changing the anion to cation ratio in a crystal were described as follows:

1. *Interpolation.* In interpolation, extra atoms are introduced into the structure in positions that are normally unoccupied in the parent phase. The defects relevant to interpolation are *interstitials*.
2. *Subtraction.* Subtraction simply means that some of the atoms that should be present in the structure are missing. The defects involved are *vacancies*.
3. *Substitution.* In this case, atoms of one type are substituted for those of another type in the structure. There is no one sort of defect associated with substitution, as the nature of the substitution will control the nature of the compensating entities required.

The structural information given in the later chapters shows that ordered or disordered aggregates of defects are the rule rather than the exception in non-stoichiometric crystals. The two extremes can be brought together by using two important variables; the interactions between the defects and the structure of the non-stoichiometric phase. The interaction between the defects can be thought of in terms of free energies or in terms of the balance between enthalpy and entropy. Random arrangements of defects implies a high entropy and weak interactions, while ordered arrays of defects means that the entropy contribution is small, the enthalpy of the interactions is high and the phase is likely to be a stoichiometric compound.

The interaction between these two parameters is shown in Table 11.1. On the left of the diagram the situation in normal stoichiometric materials is considered. In such compounds, Frenkel and Schottky defects are found. The interaction between these defects is relatively weak. Entropy effects are dominant and enthalpy effects are negligible in this case.

As we move to the right we come to non-stoichiometric phases with vacancies, interstitials or substituted atoms. If the interactions between these defects are still weak, then they will be distributed at random and entropy will still dominate enthalpy. It is very uncertain whether such a situation will

Table 11.1 A summary of defect organization and structure in non-stoichoimetric phases

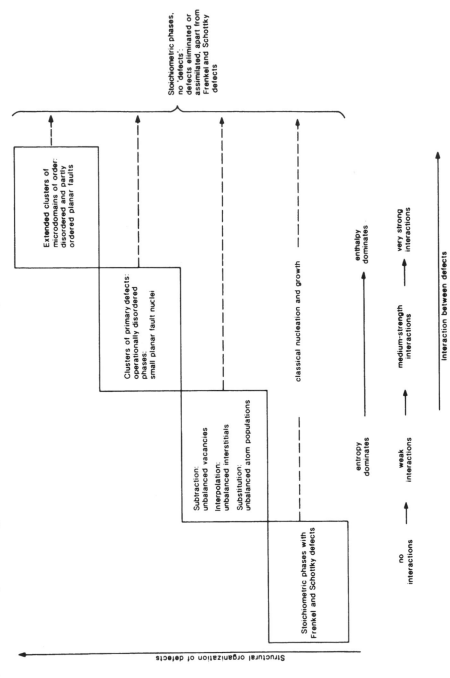

exist in any real compound, as defect interactions will certainly become dominant at very low defect concentrations. In this case, the structural situation will demand clusters of defects and this level of association is shown in the next column of the table. The nature of these clusters will depend upon the system under consideration and may contain, for example, small crystallographic shear plane nuclei. The phase will be non-stoichiometric in an operational sense, which means that experimentally the material will behave as if it contained point-defect populations. However, a degree of ordering at a microscopic level, much beyond this model, is really occurring. As the interactions between the defects increase, so the level of organization between the defect clusters will increase. The structural picture will now be of *microdomains of order* within the crystal matrix. At this juncture enthalpy will dominate entropy.

If one imagines the interactions to be so great that the defects become completely ordered, then they will be *totally assimilated* into the structure and no defects as such will be present. Thus, fully ordered Ti_5O_9 will be a defect-free stoichiometric phase. It may contain either Frenkel or Schottky defects, of course, but these will not change the composition at all. We are again back at the same situation that we encountered on the left of the table and so could imagine the sheet to be wrapped around into a cylinder.

Although such a scheme provides an attractive summary of possibilities, it is unlikely that the sequence from left to right will be followed by any one material. If the interactions between defects are weak, then only the left side of the chart will be of relevance. If the interactions are strong, then we will pass directly to microdomains or to a new ordered phase.

Temperature will also have an important effect. An increase of temperature will tend to decrease interactions and be equivalent to emphasizing the entropy factor against the enthalpy. The scheme outlined in Table 11.1 will change as the temperature increases, with the lower left-hand corner growing at the expense of the upper right-hand ordered region.

11.2 Thermodynamics and structures

The previous section relies upon experimental results. Just how easy is it to characterize a material as being non-stoichiometric at all temperatures? Structural studies such as X-ray diffraction and electron microscopy are usually carried out at room temperature. The defects present under these conditions might not be present at all temperatures. On the other hand, thermodynamic measurements of the way in which composition varies with partial pressure of the components present are made at high temperatures. In principle, therefore, it should be possible to resolve the problem of a change of defect type with temperature by combining these two techniques.

What are the problems that are encountered in practice when this is attempted? In Figure 11.1 we show the way in which the composition of a sample of $Tb_{11}O_{20}$, that is, $TbO_{1.8182}$, changes with temperature at a fixed oxygen pressure. An examination of the *reduction path* seems to indicate that we have two stoichiometric phases, $Tb_{11}O_{20}$ and Tb_7O_{12}, and that the Tb_2O_3 phase is non-stoichiometric and oxygen rich, with a composition of $TbO_{1.5+x}$. On *reoxidation*, however, quite different behaviour is found which is not so easily interpreted. In addition many regions of the curves are neither horizontal nor vertical, which indicates bivariant behaviour quite at variance with the reduction cycle. The problem is compounded by the fact that these curves are quite reproducible, and so cannot be dismissed as indicating that a non-equilibrium situation holds.

Now suppose that we have some structurally complex phases present, as the formulae $Tb_{11}O_{20}$ and Tb_7O_{12} suggest and that these phases contain differing numbers of ordered 'defects' in the parent TbO_2 phase. Reduction, which involves putting in more 'defects', will clearly require a different mechanism than oxidation, which will involve removal of the 'defects'. There is no *a priori* reason why these two processes should take place at the same rate and so, in cases involving a series of microphases, *hysteresis*, as shown on Figure 11.1, would be expected to be the rule. Any form of structural analysis at the temperature of the experiments would be invaluable.

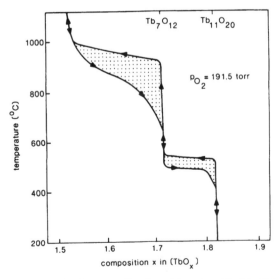

Figure 11.1 An oxidation–reduction curve for the oxide system Tb–O obtained at a constant oxygen pressure of 2.25×10^4 Pa (191.5 Torr). The arrows indicate the paths followed during oxidation and reduction, which are reproducible and not coincident. [Data reproduced from B.G. Hyde and L. Eyring, *Rare-earth Research*, Vol. 3, ed. L. Eyring, Gordon and Breach, New York (1965).]

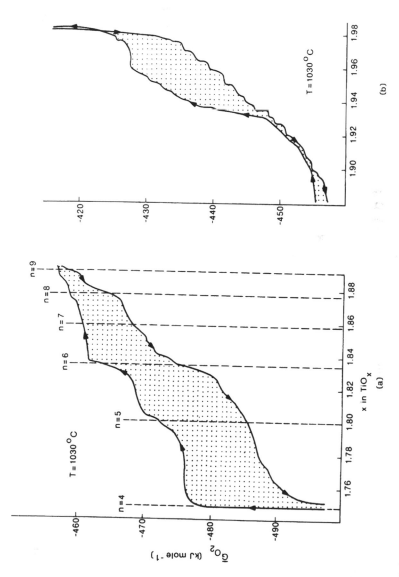

Figure 11.2 Oxidation and reduction curves for rutile, TiO_2, at $1030\,°C$. The dashed lines in the left of figure (a) show the compositions of the Ti_nO_{2n-1} phases. The region where the curves are close to coincidence in figure (b) is in the composition range where the CS planes change from $\{121\}$ type to $\{132\}$ type. [Data reproduced from R.R.Merritt and B.G.Hyde, *Philos. Trans. Roy. Soc. London* **A274**, 627 (1973).]

We can consider this in a little more detail. In Figure 11.2 we reproduce some very accurate oxidation and reduction data for the rutile form of TiO_2. It is clear that there is considerable hysteresis and it is very difficult to be precise about the number of phases present in the composition range spanned, let alone whether the behaviour at any one point should be classified as univariant or bivariant. Fortunately, the structures of the phases occurring in the system are well known. The composition range between TiO_2 and Ti_4O_7 is spanned by a series of crystallographic shear phases with a series formula of Ti_nO_{2n-1}. In the lower composition region, the structures of Ti_4O_7, Ti_5O_9, Ti_6O_{11}, Ti_7O_{13}, Ti_8O_{15} and Ti_9O_{17} all contain ordered arrays of crystallographic shear planes on $\{121\}$ planes. In the composition range between $Ti_{16}O_{31}$ and TiO_2, the crystallographic shear planes lie upon $\{132\}$ planes. Between these two regions, at a composition of about $TiO_{1.91}$, the crystallographic shear planes swing from one orientation to the other to form an infinitely adaptive phase range.

The process of introducing and ordering planar boundaries will undoubtedly be quite different to the process of removing and reordering the remaining planar boundaries. Hence, it is hardly surprising that interpretation of the thermodynamic data is so difficult. Indeed, the thermodynamic data cannot be interpreted satisfactorily without the structural information also being available.

Therefore, the problem of interpretation of thermodynamic data is not dissimilar to the problem of interpretation of structural data. The precision of the interpretation will depend upon the precision of the technique. In a system containing a homologous series of compounds, it may be impossible to differentiate, in practical terms, between a bivariant region and a closely spaced series of univariant equilibria using thermodynamic means. Figure 11.3

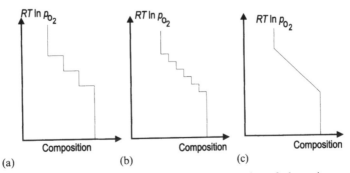

Figure 11.3 An idealized diagram showing that as the number of phases in a composition range increases, the expected free energy versus composition curves approach the continuous slope expected from a non-stoichiometric phase. The ability of the experiment to differentiate between the situations shown in (b) and (c) is of the greatest importance in practice.

shows this schematically. Similarly, it is not always easy to distinguish between a non-stoichiometric compound and a homologous series of phases using powder X-ray diffraction.

Modulated structures pose particular problems as they do not fit into any precise structural or thermodynamic categories. Further work is needed to clarify this area of concern.

11.3 Theories and calculations

An ideal way of linking the theoretical ideas about non-stoichiometric compounds with the experimental determination of defect structures and microstructures is to evaluate the theoretical predictions numerically. This is now possible because of the rapid increase in the power of computers to carry out large numbers of arithmetical calculations in reasonably short time spans. There are a number of areas in which such calculations perform an invaluable service. Of prime importance is the calculation of data that are not available experimentally, for example, if we wish to determine how crystal structures change in the high temperatures and pressures far under the surface of the Earth, for the purposes of earthquake prediction. Computation can also tell us something about dynamic processes in materials. It is not possible to follow the diffusion of single ions directly, only to evaluate the statistical result of many diffusion steps of many ions. Calculations can allow us to determine the pathways followed by individual particles. Of obvious importance are calculations which pertain to conditions which are too dangerous for experiments. Thus, calculations of the effects of fission products on solids are useful in providing information where none exists and where experiments are not possible. In fact, the list of potential applications extends into all aspects of solid state chemistry and physics. Here we will focus on what calculations have taught us about non-stoichiometry and defects. The precise details of how to perform the calculations and the details of the methods used can be found described in several books and review articles listed in the supplementary reading section.

The methods used to calculate the properties of interest follow two routes. The first involves solving the Schrödinger equation for the system of interest. These calculations are usually referred to as *quantum mechanical* or *electronic structure* calculations. There are several inherent problems with this method. The most important is that it is not possible to solve the Schrödinger equation exactly for any multi-atom system. The technique relies upon certain degrees of simplification and approximation to arrive at solutions which can be evaluated numerically. Depending upon the way in which these equations are then processed, the resulting calculations are referred to by two different names. *Semi-empirical* methods use experimental data to make estimates of some of the quantities in the equations. *Ab initio* methods avoid the use of experimental data and calculate everything from fundamental constants.

Because of the complexity of the calculations only relatively few atoms can be included if the calculation times are not to become excessively long. The method of calculation used to overcome this is also approached in two ways. The most obvious is to calculate the electronic structure for a *cluster* of atoms. The positions of the atoms can then be varied in a systematic way and the configuration corresponding to the minimum energy is taken to be the stable state in nature. The calculations give the electron density distribution, information about the bonding between the atoms and inter-atomic potentials. The chemical and physical difficulty with the method lies in selecting a realistic cluster geometry at the outset and in somehow accounting for the unformed chemical bonds at the edges of the clusters. As computer speeds increase, cluster sizes can be increased and the last limitation will decrease in importance. The alternative method is to choose a unit cell in which the boundary problems are eliminated by a computational method in which the atoms at one side of the unit cell are linked with the atoms at the corresponding other side, so as to eliminate the cell edges. This technique has been used for many years under the name of the *periodic boundary condition* method. Although this allows us to treat a unit cell of material rather than a cluster, the number of atoms contained in the cell still needs to be limited.

The quantum mechanical approach, despite the inherent difficulties in the method, gives good results for the electronic properties of materials. The band structure of a solid is invariably calculated using these techniques and recently it has been used to calculate inter-atomic potentials, of central importance in the second method of theoretically exploring the structures of solids.

The general name for this second approach is *simulation*. In this set of techniques, inter-atomic potentials are used to for the purposes of calculations. These inter-atomic potentials are generally defined between pairs of atoms and are written down as a mathematical function of the positions of the two atoms involved. Having set up the inter-atomic potentials, the positions of the nuclei are varied and the variation of the total energy as a function of atomic position is calculated. Simulation techniques have reached a high degree of sophistication and have been used to calculate the effects of shock waves on solids and the way in which cracks can be propagated through materials as well as the defect structures of interest here. The major problem encountered in these calculations lies in the accuracy of the inter-atomic potentials used. There are two different approaches to this problem. One is to use the quantum mechanical methods outlined above to calculate inter-atomic potentials that can be used in simulation calculations. The other widely used method is to estimate inter-atomic potentials from physical properties such as elasticity.

The results of simulation computations fall into three broad classes. The first of these can be called *energy minimization*. The positions of the atoms or defects in a structure of interest is varied in a systematic way and the energy calculated for each arrangement. The minimum energy structure is regarded as the one of importance in nature. In this way, one is able to determine details of surface structures and which defect configurations are most likely to be found. The energy minimization technique can also be used to discriminate between the different sites available for molecular absorption and so is being widely used to understand the processes taking place during catalysis.

A drawback of the last method is that it is a static method in which the atoms are moved in increments and a result calculated on the new static arrangement. An exciting technique uses *molecular dynamics* calculations which have become possible with increasing computational speed. This is also a simulation technique using inter-atomic potentials, but the equations evaluated include the kinetic energy of the system which considers the trajectories of some of the atoms in the structure. This technique allows the paths of diffusing ions to be constructed and so allows one to directly visualize ionic diffusion. At present, the large numbers of calculations involved allow only about 1 ns of real time to be 'visualized', but as computers improve this will increase.

The final use for simulations involves *Monte Carlo methods*. In this technique, the atoms in an array are moved small distances at random and the new configuration is accepted or rejected depending on some chemical or physical criterion, often the potential energy of the system. The rejection is not a simple yes or no affair in these calculations but varies with the degree of departure of the system from that chosen. For example, if the potential energy is used, a configuration with a slightly higher potential energy than the minimum is accepted as being highly likely to occur, while if the energy is far from the minimum it will be rather unlikely to form. The technique is essentially statistical in nature and, therefore, is able to give results concerning the likely configuration of defects in highly doped systems or information about how these defect populations are likely to change with temperature.

11.4 Defect structures and configurations

Some of the earliest simulation studies were aimed at clarifying the defect structures of technologically important materials. To illustrate the sort of results which can be obtained by these simulations, we can look at the defect

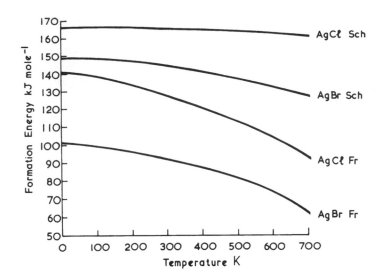

Figure 11.4 The variation of the formation energy of Frenkel and Schottky defects in AgCl and AgBr as a function of temperature. [Original data from C.R.A. Catlow, *MRS Bull*. **XIV**, 23 (1989).]

structures of those important photographic chemicals AgCl and AgBr. Even the first energy minimization calculations on these systems indicated that the formation of cation Frenkel defects required a lower energy than the formation of Schottky defects. More recent calculations have successfully reproduced the lattice parameter of the silver halide material, confirming that the cation Frenkel defects are more favoured and giving details of how the defect formation energy varies with temperature. Some of the results are plotted in Figure 11.4. Many other examples of defect structure calculations will be found in the supplementary reading listed in section 11.7.

11.5 Surfaces and interfaces

The calculation of surface and interface energies has an importance in the areas of corrosion, reactivity and phase equilibria studies. As an example, we can look at the stability of some layered perovskite phases, $Sr_{n+1}Ti_nO_{3n+1}$, illustrated in Figure 11.5. Early studies showed that only the phases Sr_2TiO_4, $Sr_3Ti_2O_7$ and $SrTiO_3$ were usually found experimentally and that $Sr_3Ti_2O_7$ coexisted with $SrTiO_3$ in samples which had a composition corresponding to n greater than 2. Calculations of the lattice energy of these phases revealed that the members of the series in which n was greater than 2 were unstable with respect to disproportionation into $Sr_3Ti_2O_7$ and $SrTiO_3$ thus

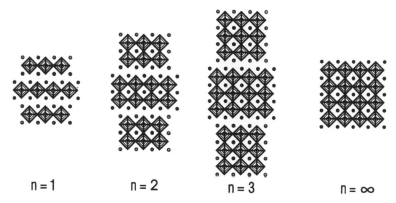

$$n = 1 \qquad n = 2 \qquad n = 3 \qquad n = \infty$$

Figure 11.5   The idealized structures of the $Sr_{n+1}Ti_nO_{3n+1}$ oxides with n taking values of 1, 2, 3 and ∞ ($SrTiO_3$). Calculations show that those structures with $n > 2$ disproportionate into $SrTiO_3$ and $Sr_3Ti_2O_7$.

$$Sr_4Ti_3O_{10} \longrightarrow Sr_3Ti_2O_7 + SrTiO_3$$

In addition, calculation of the interfacial energy between the $Sr_3Ti_2O_7$ and $SrTiO_3$ indicated that $\{100\}$ interfaces had the lowest interfacial energy and were to be preferred. Again the calculations are in good agreement with experimental results. The calculations also suggest that it is energetically preferable for the $SrTiO_3$ structure to accommodate a small excess of SrO, in the form of a thin lamellae of $Sr_3Ti_2O_7$, rather than as other defects or defect clusters. Calculations, therefore, have vindicated the observation that point defects are not used to incorporate SrO in $SrTiO_3$ and explains the limited range of the series $Sr_{n+1}Ti_nO_{3n+1}$ as well as giving valuable information about the interface between the phases.

11.6 Molecular dynamics

Experimentally, diffusion cannot be studied by recording the motion of just one atom, although this might be extremely useful. Simulations can, however, do just this. By using molecular dynamics, the low energy trajectories of atoms can be worked out and the diffusion mechanism elucidated at an atomic level. We will consider two examples which show that diffusion can be much more complicated in practice than indicated in chapters 2 and 3.

The case of Li_3N has already been mentioned in chapter 5. It was pointed out that the rapid migration of Li vacancies is the root cause of the high ionic conductivity registered by this material. However, this does not explain why the conductivity is so high. Dynamic calculations have shown that the migration is not a series of single atomic jumps but that strings of atoms move in a correlated fashion. An example is shown in Figure 11.6. Here we see that a chain of six Li^+ ions all move together so as to displace the vacancy by seven places in one movement. Such *correlated motion* increases the observed diffusion coefficient enormously.

A second example concerns diffusion in the *fluorite* structure material, $RbBiF_4$. In this material, the Rb and Bi atoms are distributed at random over the metal atom positions. Fast ion conduction is due to the migration of anion Frenkel defects formed when F^- ions, in their normal lattice positions at the corners of cubes of anions, are displaced into interstitial sites at the cube centres. It has been suspected that the diffusion of these F^- ions takes place by way of an *interstitialcy* mechanism, where the interstitial F^- displaces a neighbouring F^- on a normal lattice site into an adjacent interstitial position. Molecular dynamic simulations reveal not only that this mechanism is correct, but that the path is angled rather than straight, as shown in Figure 11.7. The tangled line shows the path of an individual F^- interstitial ion. Starting in the bottom cube, the line shows that the interstitial is moving about in the interstitial site but never moving far. Eventually, it is able to jump to the adjoining normal site, knocking the ion into the interstitial site in the upper cube. In fact, calculations have shown that the motions of more than just three atoms are involved and that the correlated motion of several F^- ions, over normal and interstitial sites, are responsible for the high ionic conductivity in this compound.

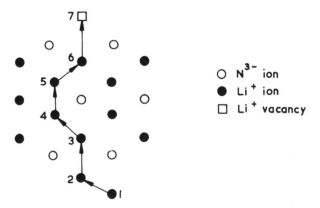

Figure 11.6 A schematic illustration of the correlated migration of six Li^+ ions in Li_3N to move the vacancy from position 7 to position 1. [Redrawn from C.R.A. Catlow, *J. Chem. Soc. Faraday Trans.* **86**, 1167 (1990).]

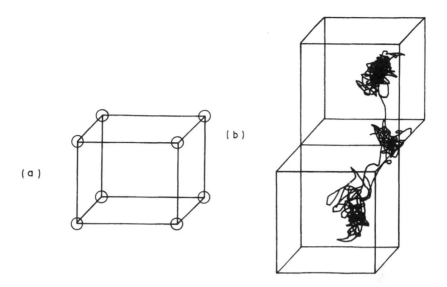

Figure 11.7 The calculated interstitialcy motion of a migrating F^- ion. (a) The basic F^- cube making up the structure. (b) The path, which looks like a tangle, shows an interstitial vibrating about its position at the centre of a cube of anions in the lower part of the figure. Eventually, it is able to make a jump to a normal anion site at a cube corner. Again it vibrates, indicated by the tangled track, until it is able to make a second jump into the new interstitial position in the top cube. Overall, there is a non-linear track for the motion. [Redrawn from C.R.A. Catlow, *J. Chem. Soc. Faraday Trans.* **86**, 1167 (1990).]

11.7 Supplementary reading

The following references cover significant or recent advances in the study of non-stoichiometric compounds, especially the relationships between thermodynamics and structure. Most work is concerned with oxide chemistry which reflects the current situation in this area of study.

Two books which contain a collection of advanced review articles are:

E. Rabenau (ed.), *Problems of Non-stoichiometry*, North–Holland, Amsterdam (1970).
O.T. Sørensen (ed.), *Non-stoichiometric Oxides*, Academic Press, New York (1981).

The whole topic is reviewed succinctly by:

D.J.M. Bevan, *Comprehensive Inorganic Chemistry*, chapter 49, Vol. 4, ed. A.F. Trotman-Dickenson, Pergamon, Oxford (1973).

The relationships between structure and thermodynamics are set out clearly by:

J.S. Anderson in: R.S. Roth and S.J. Schneider (eds.), *Solid State Chemistry*, N.B.S. Spec. Pub. 364, National Bureau of Standards, Washington (1972).
C.N.R. Rao (ed.), *The Chemistry of the Solid State*, Marcel Decker, New York (1974).
E. Rabenau (ed.), *Problems in Non-stoichiometry*, North–Holland, Amsterdam (1970) p.1.

Statistical thermodynamic theories are discussed by:

L. Manes, *Non-stoichiometric Oxides*, ed. O.T. Sorensen, Academic Press, New York (1981).

There are many reviews and books which are concerned with the calculation of the normal and defect properties of solids. Some of these of most relevance to this chapter, which also reveal the rapid evolution of the subject, are:

C.R.A. Catlow, *Solid State Chemistry Techniques*, chapter 7, eds. A.K. Cheetham and P. Day, Clarendon Press, Oxford (1987).
C.R.A. Catlow, *MRS Bull.* **XIV**, 23 (1989).
C.R.A. Catlow and G.D. Price, *Nature* **347**, 243 (1990).
C.R.A. Catlow, J.D. Gale and R.W. Grimes, *J. Solid State Chem.* **106**, 13 (1993).
P.A. Cox. *Chemistry in Britain*, March, 44 (1997).
C.R.A. Catlow (ed.), *Computer Modeling in Inorganic Crystallography*, Academic Press, New York (1997).

Formula index

Index of structures

Subject index

Printed and bound by CPI Group (UK) Ltd, Croydon, CR0 4YY

22/10/2024

01777622-0013